AIRBORNE FOR PLEASURE

ALSO FROM DAVID & CHARLES

The Aviator's World, by Michael Edwards
Instruments of Flight, by Mervyn Siberry

AIRBORNE FOR PLEASURE

A Guide
to Flying, Gliding, Ballooning
and Parachuting

ALBERT MORGAN

DAVID & CHARLES

NEWTON ABBOT   LONDON

NORTH POMFRET (VT)   VANCOUVER

To my wife Betty, and daughters Susan and Janice, who managed not only to hold their breath during my aerial researches but also, more incredibly, their tongues

ISBN 0 7153 6477 4

LOC 74 20449

© Albert Morgan 1975

Set in 11 on 13pt Garamond and printed in Great Britain by Latimer Trend & Company Ltd Plymouth for David & Charles (Holdings) Limited South Devon House Newton Abbot Devon

Published in the United States of America by David & Charles Inc North Pomfret Vermont 05053 USA

Published in Canada by Douglas David & Charles Limited 132 Philip Avenue North Vancouver BC

CONTENTS

LIST OF ILLUSTRATIONS

INTRODUCTION

For centuries Britain's historical associations have been with the sea, and in their leisure activities Britons have tended to follow marine traditions on coastal waters and rivers. But today leisure in the air is thriving, despite the controls imposed by the tremendous growth in commercial flying.

Only forty years ago the airspace above us was completely free. When Imperial Airways introduced its luxurious Handley Page 42 on the London–Paris service in 1931, its mid-day departure from Croydon was probably the only scheduled flight over Britain at that particular time of day. The pilot would head south to the familiar shoulder of the North Downs at Redhill, turn east and pick up the London–Folkestone railway line down to the coast. Stations such as Tonbridge and Ashford had their names boldly painted on the roofs for the benefit of aviators. Passengers were supplied with maps and invited 'to see the whole glorious panorama of Surrey and Kent spread out beneath your eyes till you cross the silver thread of the Channel'. The aircraft flew at a sedate 98mph for a 2hr 20min flight to Paris which gave the thirty-eight passengers ample time to enjoy the pleasures of flying at leisure. That was forty years ago, and such pleasures are still available today for those who care to reach out for them.

A false impression persists today that airspace above the United Kingdom is one vast control zone. True, the official airlanes slice across the sky, carving great flight paths sometimes ten miles wide, but unless commercial and military aircraft are making their approaches to airports or aerodromes they are normally far above the modest demands made on airspace by the light aeroplane, the glider and the balloon.

Certainly freedom in the air is not as before, but in most areas the private pilot is still free to take-off from his own landing strip, sometimes within fifteen miles of central London, without being required to seek permission from any authority whatsoever. Further, once equipped with radio and certain licence qualifications a pilot may use airlanes, under Air Traffic Control, and land at an international airport, although that particular welcome would not be a warm one. The National Air Traffic Services, which monitors by radar all airspace over the United Kingdom, is quite prepared to assist all aircraft right down to the glider, once the pilot has the means of communication to enable him to contact the service.

The third dimension gives an entirely new aspect to the countryside. The hill seen merely as a ridge at ground level often takes on more detail from the air. It is no longer just a hill but perhaps a perfect semi-circle or horseshoe, with only grass and shrub on the outside escarpment but with a fine inner lining of young trees protected by the brow. Motor through East Anglia and the land looks quiet and flat, mile after mile. See it from the air and the ground is cut by the herring bones of the land drainage system, running into the sharp meanderings of the Yare, the Ouse and the Waveney.

A very experienced flying instructor once told me that learning to fly is 90 per cent psychological and 10 per cent technical, and that in the early stages thought processes operate at only something like 30 to 50 per cent of normal capacity. The balance is restored by accepting the third dimension, overcoming the inborn fear of falling, and getting the body to accept the changing force of gravity, never encountered in ordinary life other than on the occasional humped-back bridge. These very human fears are quite normal and once understood and accepted, the novice flyer can soon become 'airminded'.

This book is in no way a textbook. Other writers are far better qualified to deal with specific training. Instead, my researches have been on behalf of the potential newcomer to air leisure seeking an answer to the basic question: 'Is this for me'? The reader can consider the financial commitments involved in becoming proficient, the depth of study demanded by the stringent regulations, and the important point of how far one would need to travel for training. For this purpose, a complete list of clubs and schools is included in the reference section.

Aptitude and just plain nerve are very personal factors, but a good measure of self-examination can be derived from the descriptions of training requirements, without the reader having to spend considerable time questioning a busy club secretary item by item. Not every individual is prepared to respond to the breezy 'Come along and find out' attitude.

Rightly and necessarily, all training is in accordance with a strict syllabus and, to emphasise this, considerable space has here been given to the requirements dictated by government regulation, as well as those self-imposed by the various associations. Cost illustrations can be no more than general, but the reader will find some guide here as to what his pound will buy. Manufacturers and clubs will be only too pleased to quote specific prices and fees on application.

CHAPTER ONE

FLYING FOR FUN

THE PRIVATE OWNER

The ideal situation for a private pilot today is to be the sole owner of his light aircraft, with business use to offset the costs and sufficient land to lay out an airstrip adjoining his home. This is not as uncommon as might first appear, though ownership of a private aircraft is an expensive matter and using it for anything less than 250 hours flying a year may be considered entirely uneconomic. Even operating at 250 hours annually calls for considerable use—averaging five hours a week and upwards of 500 miles flown—and if allowance is made for adverse weather to upset the average, this is a high utilisation indeed. Economic factors, however, are apt to be readily discounted where a favourite sport or pastime is concerned and if motorists were to make an estimate of the rocketing cost of owning a car their first consideration, there would be far fewer privately owned vehicles on the roads today.

Once the would-be private owner has rationalised the costs of his flying, the type of aircraft he selects will depend largely on the depth of his purse. Few people have suitable land on which to develop an airstrip—though it might well be, for example, that two or three pilot farmers, perhaps having consultancies in agriculture which require travel to many parts of the country, could work out a satisfactory arrangement to share an airstrip. Hangarage or covered parking areas are important and the aircraft insurers would be particularly interested in this factor. Different types of aircraft need different runway lengths and while a four-seater may get off the ground quite safely with one passenger, the presence of a full load could present problems. A small grass strip adequate for one type of light aircraft may prove useless should the owner decide to go for a bigger machine.

Owners of landing strips tend not to advertise their existence for three main reasons. First, the strip is for private use, and flying activity sometimes brings spectators, their cars and allied safety problems; second, damage to fences and crops by thoughtless spectators attempting to get closer views of take-offs and landings; third, the threat of vandalism and damage to aircraft after dark.

Many private airstrip users set their own code of conduct. That is no 'circuit bashing', but only straightforward up-and-away flying to minimise inconvenience to property-owners and residents. Fuel supply regulations usually preclude the storage of petrol in any quantity and oblige private owners to fly to the nearest aerodrome for fuelling. This presents little difficulty since there are more than 170 aerodromes in the United Kingdom, State-owned or licensed by the Civil Air Authority, serving civil aircraft.

There are also unlicensed airfields where operating facilities may be obtained by permission of the owner or controlling body. If one adds the innumerable private landing strips where the owner may grant facilities for an occasional landing by prior arrangement over the telephone, the private pilot will be seen to have access to a vast number of suitable landing grounds, for business or pleasure.

The major airports such as those at Heathrow, Gatwick, Prestwick and Edinburgh, are administered by the British Airways Authority. Although light aircraft are not legally barred from using these international airports, they are not encouraged and, in turn, the private pilot is not normally concerned with landing on them. Other licensed aerodromes are operated by local municipal authorities, while certain military airfields may be used by arrangement. These aerodromes obviously vary considerably in landing aids and technical equipment, ranging from the basic radio advisory watch to the more sophisticated airport control procedure.

For the private pilot, then, there is a fair network of aerodrome facilities and, unless he is operating an ultra-light aircraft under the special 'Permit to fly' scheme, where certain maintenance operations may be carried out by the pilot himself, the demands of the Certificate of Airworthiness will necessitate periodic visits to aeronautical engineers' workshops based on certain aerodromes in each area. So we turn from the ideal situation of a runway at the bottom of the garden to the more realistic one of keeping an aircraft at an aerodrome. Driving any distance to get there may take the edge off the day, but the problems involved by wind strength and direction are more easily resolved at an aerodrome than when operating from a 50ft wide landing strip. Engineering facilities will also be close at hand, with the added benefits of club membership along with other flying enthusiasts.

OPERATING COSTS

For aerodrome-based aircraft, it is possible to give a general assessment of costs. Dealers can—and do—argue all day about the virtues of their own particular model and respective running costs, but here is a broad picture of the expense facing the would-be buyer of a single-engined, four-seater Piper Cherokee at around £11,500—plus the cost of certain navigational and radio aids.

The expense of owning an aeroplane can be divided into fixed costs and running costs. The fixed costs are parking, or better still, covered hangarage, insurance, maintenance and depreciation. These charges arise whether the aircraft flies or not.

First, hangarage. Floor space rentals on aerodromes have kept pace with inflationary trends as much as any other properties. While it may still be possible to close park an aircraft in a crowded hangar for a monthly rental of £20, most owners have to face much higher rents, and £150 per annum is merely a starting price. Insurance is dealt with more fully in a later chapter, but the owner may need to set aside some 4–5 per cent of the hull value for this item. Incidentally, the hull may be defined as the aircraft without radio, navigational aids and other accessories. Comprehensive insurance may cost, in this instance £500 per annum. All British-registered aircraft have to be maintained to the standards required by the government-specified Certificate of Airworthiness and any necessary repairs carried out by an authorised engineer. Renewal of the certificate takes place every two years, when a detailed inspection of the aircraft and its engine is called for. This may cost £350 on top of any repair or routine work, so at least £175 a year must be set aside under this heading. An aeroplane must also be examined by an authorised engineer after every fifty flying hours, or every four months, whichever comes first. This check may cost about £35–£40. So much for the fixed costs—in this illustration £1,000 a year, with a heavy plus sign for parts and any additional labour costs.

Once airborne, the Cherokee—which may be considered a four-seater family saloon of the air—will consume seven gallons an hour and cruise between 120 and 130mph according to engine type. Fuel costs are equivalent to top-grade petrol for motoring but should the aircraft be flown to an overseas destination there is a fuel rebate of 66 per cent, that is one-third the UK rate, on the fuel that the machine has in its tanks on leaving the UK aerodrome of departure. There is no annual limit on this drawback and it can be used on day trips to the Continent if desired.

The engine of a Cherokee must be overhauled every 2,400 hours, and this at current prices would cost approximately £1,400. The total of 2,400 hours is a long time in the life of any aircraft which, broadly speaking, is as new as its last 50-hour check, and even at the high utilisation rate of 240 hours a year would obviously take ten years to come round. Sales agents suggest that 60–65p an hour should be set aside for this engine replacement.

On the aspect of depreciation, an aircraft's market value depends very much on its stage of airworthiness at any given time, though importers of American models in particular point out that, what with price increases and currency fluctuations, the value of second-hand models holds up sufficiently well virtually to rule out depreciation. This is one point of view, although in reality to replace an aircraft is still going to call for additional cash, however one tackles the sums.

In the regrettable absence of a comparable British-built model,

Page 17 (*above*) A popular type of light plane for private ownership or training, the French-built Rallye Club 100 3–4-seater cruises at about 100mph with a 100hp engine; (*below*) David Fairclough (foreground), CFI of the London School of Flying, gives the author a pre-flight cockpit checkout in a Piper Cherokee

Page 18 The American-built Bell 47 helicopter, a three-seater with a cruising speed of about 90 mph at a height of 5,000ft

let us take a look at the table overleaf provided by Air
Touring Services of Biggin Hill, who are agents for Aerospatiale
General Aviation, a subsidiary of the French builders of Con-
corde, and the German Messerschmitt-Bolkow-Blohm group.

The table relates to the French Rallye range of light aircraft, of
which there are now over 2,000 flying in various parts of the
world. The first column, 100 S, relates to the Rallye Sport 100,
a full three-seater, or two-plus-two to the motoring world, with
a cruising speed of 94mph at 75 per cent power. The engine has a
1,800-hour life. Price: £6,750. (Picture, p 17.) The MS 892A
Rallye Commodore 150 is a full four-seater with a cruising speed
of 108mph and an engine life of 2,000 hours. Price: £10,200. The
MS 894A Rallye Minerva 220 is also a full four-seater with a
cruising speed of 134mph, and an engine life of 2,000 hours. Price:
£13,000. In the table, the hours flown per year envisage an ex-
tremely high utilisation, one more likely to be achieved by an
ownership syndicate than an individual owner or a partnership.

SYNDICATE OWNERSHIP

If a would-be owner is undaunted by these expensive but nevertheless realistic figures, well and good, but for those brought
down to earth with a bump there is still encouragement to be
found in the prospect of partnership or syndicate ownership. A
partnership of two people sharing the purchase of an aircraft and
its attendant fixed costs should not be too difficult to organise,
but joint ownership by possibly up to a maximum of five people
requires a long, hard look at oneself and the other personalities
involved. Obviously the object is to aim for a cross-section of
people from varying occupations to give the maximum spread of
the demands on the aircraft. Five people working the same shift
or hours five days a week may end up at best queuing for the
aircraft at weekends or evenings during the summer months. The
scheme would be doomed from the start. The mix, therefore, calls
for a good deal of commonsense and a large measure of tolerance,
and the more members free to utilise the aeroplane in midweek the
better.

TYPICAL RUNNING COSTS OF AIRCRAFT
WITH AIR TOURING SERVICES FIXED PRICE MAINTENANCE

FIXED ANNUAL COSTS (1)	(100 h.p.) 100S 2 seat		(150 h.p.) MS.892A 4 seat		(220 h.p.) MS.894A 4 seat	
Depreciation over 7 years with 30% residual	£675		£1020		£1300	
Insurance @ 3% (Private) Hull 4% (Club)	£202		£306		£390	
Parking	£120		£120		£120	
Annual provision for 2 year C. of A. survey £90	£90		£90		£90	
Miscellaneous	£60		£60		£75	
	£1147		£1596		£1975	
HOURLY COSTS (2)						
Fuel consumption (normal cruise) 4½ p.h. @ 55p 6½ p.h. @ 55p 8 p.h. @ 55p	£2.47		£3.57		£4.40	
Oil 1 pint/hour @ 88p per gallon	.11		.11		.11	
Maintenance Airframe & Engine	£1.00		£1.00		£1.00	
Maintenance Radio	.10		.10		.10	
Spares — tyres etc.	.10		.15		.15	
Engine £750/2160 Rolls Royce	.34					
£1500/2400 Lycoming			.62			
£1000/1800 Franklin					.55	
	£4.12		£5.55		£6.31	
HOURS FLOWN PER YEAR	350	700	350	700	350	700
Fixed cost as (1) above per hour	£3.38	£1.64	£4.56	£2.28	£5.64	£2.82
Hourly cost as (2) above	£4.12	£4.12	£5.55	£5.55	£6.31	£6.31
COST PER HOUR TOTAL	£7.40	£5.76	£10.11	£7.83	£11.95	£9.13
Cost per mile - pence	.07	.06	.08	.06	.08	.06
Cost per seat per mile - pence	.03½	.03	.02	.01½	.02	.01½

Assuming a congenial group of four to share the initial outlay, an aviation finance house would probably arrange a contract. All fixed costs could then be shared to make the whole proposition much more attractive financially. Insurance would, of course, come higher with the syndicate to be covered as a body and the much higher utilisation taking an extra slice of the funds. With several thousands of pounds involved and considerable responsibilities both towards the aircraft and third parties, such a syndicate would certainly need a formal agreement, preferably drawn up by a solicitor. A secretary/treasurer would be needed to establish a flight log and administer the funds, and there would have to be provisions for any member wishing to withdraw and the possibility of an application to fill the gap. The British Light Aviation Centre can give guidance in this field and a potential aircraft-owning syndicate could also, with advantage, study the Dante Balloon Group's excellent constitution which is reproduced, with their permission, in Appendix F. Once a syndicate is seen to work, it has also the merits of a small club. Most important of all, perhaps, is that with adequate working capital the group is more likely to be able to afford a worthwhile machine, one that would be out of reach of a pilot attempting to go it alone.

FLIGHT REGULATIONS

Whether the would-be flier has plans for individual, half-share or group ownership of an aircraft, or simply to hire a club machine from time to time, there is one common qualification—the Private Pilot's Licence. Though the licence has certain limitations, the private pilot still has considerable freedom within the framework laid down by the authorities. The restrictions are fully detailed in Civil Air Publication (CAP) 85, obtainable from Her Majesty's Stationery Office. This states that all flights within the United Kingdom shall be made either under Visual Flight Rules (VFR) or Instrument Flight Rules (IFR). A private pilot without an instrument rating endorsement must fly under VFR.

In controlled airspace which has not been notified as subject to IFR, the regulations state that a pilot may fly under VFR by day provided he is able to remain at least one nautical mile horizontally and 1,000ft vertically away from cloud and in a flight visibility of at least five nautical miles.

Outside controlled airspace, the regulations are that a pilot '. . . may fly under VFR at and below 3,000ft above mean sea level provided he remains clear of cloud and in sight of the surface. Above 3,000ft above mean sea level, a pilot may fly under VFR provided he is able to remain at least one nautical mile horizontally and 1,000ft vertically away from cloud and in a flight visibility of at least five nautical miles.'

The Private Pilot's Licence examination does not require a study of flying by instruments but, as a safeguard, the rules quoted above ensure that the private pilot is restricted to flying in good weather and visibility. To fly in conditions less than the minima laid down, the pilot must qualify for an Instrument Meteorological Conditions rating. Requirements for this privilege demand 100 hours flying experience (60 hours in command, 30 hours on cross-country flights and a 10-hour instrument flying course).

The more advanced Instrument Rating, by which the pilot qualifies to fly solely by instruments, may or may not figure in the long-term plans of the potential pilot. This rating permits the use of specified airway lanes and other privileges, but it does call for at least 150 hours experience as pilot in command, 50 hours cross-country flying and 40 hours experience of instrument flying. In this instance, half of the instrument flying instruction may take place in a flight simulator.

Private pilots are increasingly paying more attention to the Radio/Telephony licence. It is not necessary, by law, for an aircraft to be equipped with radio although it is important in these days of busier skies for a pilot to have full communication facilities at his disposal. During tuition for the Private Pilot's Licence, the student is permitted to use radio/telephony without a licence but it is desirable for the formal qualification to operate airborne radios to be acquired as soon as possible.

Although a private pilot may fly solo at night, he must qualify for a night rating before he may carry passengers. The Night Rating calls for a minimum of 50 hours experience as pilot (25 in command) and the course itself includes 5 hours night flying and 5 hours instrument flying training. However, for the student pilot, the Instrument Meteorological Conditions, Instrument Rating and Night Rating qualifications may be put aside at this stage. Thousands of pilots fly every year in leisure activities without concerning themselves with these endorsements.

THE BRITISH LIGHT AVIATION CENTRE

Rules governing light aviation in Britain are administered by the Department of Trade and the Civil Aviation Authority. An independent body, the British Light Aviation Centre, works in close collaboration with these authorities on behalf of the private flier, whether or not he or she is a member of the centre. In a broad context the BLAC is the airborne equivalent of the Automobile Association or Royal Automobile Club, representing the interests of aircraft owners, flying clubs and flying instructors, as well as small aerodrome operators. The BLAC was formed in 1966, from a merger of the Aviation Centre of the Royal Aero Club and the Association of British Aero Clubs and Centres. Both organisations had played an active role in sporting aviation for many years and their prestige and experience have been passed on to the BLAC. The Centre has a full-time staff under a chief executive and is based at Artillery Mansions, 75 Victoria Street, London SW1H 0JD. Members are encouraged to use and visit the Centre to discuss all aspects of aviation, such as air law, maintenance, overseas touring and the technicalities of regulations. Like most clubs, the BLAC can offer its members discount arrangements with car hire companies and hotel groups.

More than 150 flying schools and clubs in the United Kingdom are now affiliated to the BLAC and its schedule of pilotage standards is accepted by the Civil Aviation Authority as the basis of an approved course for the Private Pilot's Licence. The approved course calls for a minimum of 35 hours flying—as compared with the normal requirement of 40 hours minimum—to be completed within six months and may be given only at training centres authorised by the Civil Aviation Authority. It should, however, be borne in mind that the cost of the 35 hours approved course does not necessarily mean an automatic $12\frac{1}{2}$ per cent saving over the 40 hours course. Tuition charges vary and a school operating the approved course may have higher overheads, such as more full-time instructors, more elaborate ground services and a larger fleet of aircraft.

THE STUDENT PILOT'S LICENCE

It is not legally necessary for the pilot of an aircraft to have any licence until going solo, provided that he or she is accompanied by a qualified pilot in a dual-control machine. It is under this heading that it is permissible for flying schools to give trial lessons to would-be fliers without any formal application to the authorities. Before any course is started, however, flying schools require the student to hold the Student Pilot's Licence, which is really the aviation equivalent of a provisional driving licence. This is a sensible precaution because any lessons given would have been wasted if the Student Pilot's Licence were subsequently refused on medical grounds. The medical examination, conducted by a qualified practitioner, calls for a standard of general fitness and a physical ability to handle the controls of an aircraft. The fee is set by the medical examiner himself and the minimum age for the granting of a Student Pilot's Licence is 17 years. The licence itself costs £2, on issue or renewal, and if the applicant is under 40 it will remain valid for two years. At 40 and over, the period of validity is cut to 12 months, though the authorities may, for medical or other reasons, specify a shorter period. Application forms for a Student Pilot's Licence are available from flying clubs, or from the Department of Trade TL5, Shell Mex House, Strand, London WC2.

THE PRIVATE PILOT'S LICENCE

Once issued with a Student Pilot's Licence the novice is legally permitted to fly an aeroplane solo on specific exercises in preparation for the Private Pilot's Licence. The instructor will, of course, decide just when the student is ready to fly solo. The actual test for the Private Pilot's Licence consists of flying tests and ground examinations. Once again, the minimum age is 17 years but, unlike the Student Pilot's Licence, the Private Pilot's Licence is valid for 5 years and costs £5.

These requirements met, next comes the passing of examinations designed to show competence in the air and a general knowledge of the aircraft, aviation law, navigation and meteorology. The flight examiner is not normally a ministry inspector and is more likely to be the chief flying instructor of the pupil's own flying school, provided he has been approved as an examiner by the Civil Aviation Authority, which calls for a high degree of competence and considerable experience.

Every pilot is required to keep a flying log book and this serves as evidence of the student pilot's experience in preparation for the application for the Private Pilot's Licence (Aeroplanes), to give the document its formal title. Personal flying log books submitted in support of applications for Private Pilot's Licences must be countersigned and stamped by an authority at the student's flying school. A minimum of 40 hours flying experience (35 hours on the approved course) must have been logged, with at least 10 hours as pilot in command, which means flying solo. The full test schedule is given later. Certain exemptions may be granted to qualified service pilots and to holders of licences granted by authorities abroad, though such applicants may still be required to take a special flying test to prove their competence.

The Private Pilot's Licence comprises an identification of the owner, a current medical certificate, and the various ratings and associated certificates for which the owner may have qualified during his flying activities. The aircraft rating specifies the type of aeroplane the holder is qualified to fly, in the same way that a driving licence records the classes of vehicle which the holder may drive. For private pilots the types of aircraft are divided into three groups:

(1) Group A—all types of single-engined aeroplanes for which the maximum total weight authorised does not exceed 12,500lb.

(2) Group B—all types of aeroplanes, having two or more engines, for which the maximum total weight authorised does not exceed 12,500lb.

(3) Group C—individual types of aeroplanes for which the maximum total weight authorised exceeds 12,500lb.

A licence is not issued without including at least one of the group ratings. Group A covers practically every single-engined type, for most four-seater machines would not weigh a quarter of the maximum weight, even when fully laden. Groups B and C apply to multi-engined rating, which comparatively few private fliers wish to acquire unless engaged on business flights.

LICENCE RENEWALS

A licence has to be submitted every thirteen months to an authorised examiner for a check on the holder's current flying experience, as the regulations specify a minimum requirement to maintain the licence's validity. Pilots are ranged into seven groupings; private pilots are covered by groups A, B and C, while the others concern commercial and professional pilots only. If the private pilot's log book shows insufficient experience in the respective aircraft rating category during the previous thirteen months, he will be required to take a flying test. The required amounts of experience in the various groups are as follows:

Group A

Purpose of flight	Flying single-engined aircraft below 12,500lb when the pilot does not receive remuneration for the flight.
Experience required in each 13-month period	Five hours experience as pilot in command in an aircraft of the same class as that which the pilot wishes to fly. In the case of helicopters, including at least one flight in the type he wishes to fly.

Group B

Purpose of flight	Flying multi-engined aircraft below 12,500lb when the pilot does not receive remuneration for the flight.
Experience required in each 13-month period	Five hours experience as pilot in command of an aircraft of the same class as that which the pilot wishes to fly, of which at least one flight shall have been

in a multi-engined aircraft. In the case of helicopters, including at least one flight in the type he wishes to fly.

Group C

Purpose of flight

Flying aircraft over 12,500lb when the pilot does not receive remuneration for the flight.

Experience required in each 13-month period

Five hours experience as pilot in command of an aircraft of the same class as that which he wishes to fly, of which one flight shall have been in an aircraft of the same type.

THE FLYING TEST

The flying test for a Private Pilot's Licence takes about an hour and is conducted by a CAA-appointed examiner, who will require the applicant to complete a series of manoeuvres to demonstrate his competence in most situations likely to arise in flight, and to observe certain ground procedures. The test requires that the pilot under examination should occupy the pilot-in-command seat in the aircraft, normally on the port or left-hand side, and take all command decisions.

The syllabus for Flying Tests (Aeroplanes) published by the CAA specifies that:

During the test the candidate will be assessed on airmanship, including circuit procedure, aerodrome discipline, the setting and use of the flight instruments fitted to the aircraft, pre-flight inspection, starting procedure and running up, cockpit check and vital actions.

Normally, the test will consist of a flight of approximately one hour and will cover the following: taxying and take-off; assessment of cross-wind component; take-off with cross-wind component.

Engine failure after take-off. Action in the event of fire. Straight and level flying.

Short-field landing (touch-down to be normally within 100yd of the edge of the landing area). Assessment of cross-wind com-

ponent; landing with cross-wind component. Landing without
power from a position and height selected by the examiner.

Overshoot procedure from a powered approach.

Turns of various degrees of bank with and without engine;
climbing and descending turns.

Recognition of the approach to a stall; stalling and recovery.
Spinning and recovery. (This section will not be included if candidate produces evidence that he has recovered from a spin.)

All applicants must complete these manoeuvres, but for Groups
B and C, that is with multi-engined rating, the applicant must also
show competence in:

The manoeuvres used in normal flight with one engine inoperative.

Approach and landing with one engine inoperative.

Should an applicant fail in any part of the test, he may be required
to undertake further practice before being accepted for re-test.
The flying test and technical examinations must be completed
within the same period of six months immediately preceding the
date of issue of the licence, or the candidate will have to undergo
complete re-examination.

When flight tests are necessary to renew a rating, there is a
short examination which, for Group A, is a sequence of all
manoeuvres used in normal flight, including take-off and landing,
and simulated forced landing.

Renewal for Groups B and C (multi-engined aircraft) demands
additional tests—take-off with simulated engine failure when airborne; with one engine inoperative, all manoeuvres used in normal flight; approach and landing with one engine inoperative.

THE GROUND EXAMINATION

The ground examination subjects for the Private Pilot's Licence
(Aeroplanes) are: (i) Aviation Law, and Flight Rules and Procedures; (ii) Navigation and Meteorology; (iii) Airframes and
Engines (General); (iv) Airframes and Engines (Specific Types).
For ratings in Group C only; (v) Seaplanes, Collision Regulations
(if the licence is to include a sea or amphibian class of flying
machine).

A qualified pilot serving in the armed forces is required to be examined in subject (i) and where applicable (iv and v). The holder of a Flight Navigator's Licence is not required to pass subject (ii). To qualify, candidates must score at least 70 per cent of the marks in each subject, except (v) which is 'pass' or 'fail'. Candidates who do not pass the examination at the first attempt may be referred for re-examination in particular subjects, but the complete examination must be passed within a period of twelve weeks. Failure to do so will entail complete re-examination.

The syllabus for technical examinations is set out fully in the official Civil Air Publication 53, but for further background knowledge CAP 85 is essential reading for the student. This publication is based on extracts from the *Air Pilot*, the 'blue book' of aviation, which deals with every aspect of flight information, from radio frequencies to runway and landing facilities throughout the country. The student is not expected to buy a copy of this expensive publication, which can be consulted at his flying school or club, but a detailed knowledge of its contents is required by the first part of the technical examination; that on Aviation Law and Flight Rules Procedure.

A further requirement is a familiarity with NOTAMS, which are information sheets periodically circulated for the benefit of all pilots. NOTAMS keep the pilot up to date with day to day changes and details of a more long-term nature affecting operations in the air and on the ground. NOTAMS, too, are available for study at clubs and aerodromes. Pilots may also apply to be put on the distribution list of an information circular by writing to the Aeronautical Information Service, Tolcarne Drive, Pinner, Middlesex. There is no fee involved. The service, a joint Ministry of Defence-Civil Aviation Authority operation, gathers and circulates aeronautical information, and its functions figure in some of the questions of the technical examinations.

The paper on Aviation Law covers air traffic rules and services, such as types of airspace and air traffic units, restrictions and hazards, control zones and airways. Students are not required to quote meteorological stations by name but must know the system of obtaining flight forecasts. A knowledge must be proved of the customs and public health requirements, certain sections of the Air Navigation Order 1966, and the Rules of the Air and Air Traffic Control Regulations 1966. The particular sections are listed in Civil Air Publication 53.

Navigation and Meteorology

To the novice, the terms used in this paper may at first sight appear highly technical—aeronautical charts, isogonals, heading (true, magnetic and compass) and so on—but more familiar references to latitude and longitude and ground speeds help the student to enter the subject at the shallow end. The several parts usually fall into their allotted place under the direction of the flying instructor and ground lecturer during the Private Pilot's Licence course. Weather conditions, also included in this paper, may make a simpler study.

Airframes and Engines (General)

This is an oral test in which the student is required only to answer questions relating to the particular group of aircraft he wishes to fly. The questions are of a general nature, concerning engine controls and performance, flying controls and emergency procedures such as fire in the air. The various requirements of the Certificate of Airworthiness are also referred to. By the very nature of the training course, the answers contained in this section should present no difficulty to the prepared student.

Airframes and Engines (Specific Type)
(For Group C Ratings)

As already indicated, this applies only to ratings in Group C. The paper is confined to the type of engine and aeroplane which the applicant wishes to fly. The questions themselves are of an advanced nature, relating to pressure and vacuum systems, hydraulics and electrics. Engine characteristics and turbine engines are also examined.

Seaplanes, Collision Regulations

A pilot learning to fly seaplanes or amphibians will need a good deal of seamanship as well as aeronautical knowledge. Rules of the road in open and restricted waters, beaching and slipway procedures and signals displayed by aircraft on the water are included in the training schedule for this specialist type of operation.

If the formal but necessary statutory requirements sound intimidating, at least there is no shortage of training facilities. Flying schools affiliated to the British Light Aviation Centre total more than 250 and range from Aberdeen to Plymouth, from Southend to Swansea. Some operate 40-hour courses, others have official blessing to run the approved 35-hour course, and most of them offer club facilities to their student members. The BLAC's schedule of flight training is a carefully constructed syllabus approved by the Civil Aviation Authority and many bodies, including the Guild of Air Pilots and the Royal Air Force, have contributed to its compilation. The schedule is strictly followed, yet the student will find that tuition is tailored to his personal ability to absorb instruction, while at the same time maintaining the standard of progress and quality of instruction stipulated in the BLAC *Manual of Flying*.

CHAPTER TWO

LEARNING TO FLY

From the very first lesson—which acquaints the student with the aircraft controls—the instructor will be preparing the ground for developing confidence, while avoiding confusing technicalities. Instilling check systems and procedures is a necessary foundation for the aspiring student, and emergency drills and sequences help to foster these points. During the early flights the instructor will assess the student's disposition to flying and though it is not unknown for a student to feel airsick at first the condition is usually overcome as lessons progress.

The student, through ground instruction, will be introduced to the effects of the controls on the aircraft in flight and will later be given practical instruction in the air. He will also learn about the basics of flying, landing and general manoeuvring on the ground. This period of the student's training is spread over a minimum of nine hours dual-control instruction. From this point the instructor is assessing the pupil in readiness for the first solo flight and the BLAC Manual stresses the importance of timing the decision. If the student is allowed to fly solo before he is ready the result may be a poor and perhaps unnerving flight, but if the flight is delayed unduly the result may be a deterioration in the pupil's standard of flying and a possible loss of interest.

The student who progresses satisfactorily will be permitted to fly alone after nine hours and it will be the most important moment of his flying training, if not his airborne career. He will make only one circuit of the aerodrome unless he finds it necessary to overshoot and go round again. The syllabus allows fifteen minutes for the first solo and it says much for the instructors' assessments of student pilots that, while there may be occasional bad landings during first solos, there are rarely bad accidents.

During the next seven hours of the course, the student is given tuition in consolidation exercises and circuit flying, approach and landing. This period includes some four hours of solo flying—again on the decision of the instructor. Ground instruction geared to flying exercises takes the student into advanced turning and steep turns in excess of 40 degrees, to develop co-ordination and confidence. Forced landings, with and without power, give the student practice in the selection of landing areas and the recognition of surfaces ranging from sandy beaches and marshland to grass and stubble. Further ground lectures will prepare the student for the navigation and other papers contained in the Private Pilot's Licence examination.

The 35-hour approved course requires the student to have 3 hours cross-country solo flying experience, so that by the time he has flown a 'mock' Private Pilot's Licence test, under the examination of the chief flying instructor, he will have flown a minimum of 24 hours dual control and 11 hours solo.

It should be stressed that even for a student pilot of exceptional ability a 35-hour course requires that the lessons be over as short a time as possible to ensure continuity. If he tries to spread the course to just fit inside the required six-month period he will find that he needs more lessons to keep up the required rate of progress. Clubs and schools operating 40-hour minimum courses are not limited to the six-month rule, but the student having perhaps only one hour of tuition a week may well need additional tuition to maintain a firm grasp of his studies.

WHAT IT COSTS

So we arrive at costs. Learning to fly, we are told, is not cheap, but it is difficult to find a yardstick to measure what is 'cheap' and what is 'expensive'. Fifty hours spent in a driving-school car may add up to £150 . . . a summer holiday for two in the Eastern Mediterranean may give no change from £400 or £500.

For £500 we are within range of the Private Pilot's Licence. Tuition fees, at the time of writing in 1974, vary between £13 and £16 an hour. Take the price structure operated by the London School of Flying, a Civil Aviation Authority-approved club based at Elstree Aerodrome, close by the M 1, a few miles north of London. This school concentrates on straightforward tuition rather than fostering a club-like atmosphere. A trial lesson at Elstree costs about £7.50, and is really a familiarisation flight designed to introduce the would-be flier to the general surroundings of the aeroplane's cockpit. Under close supervision the instructor lets the client take over the controls to get the feel of flying, and to give him a chance to ask himself, 'Is this for me?'. He can then go home and think about the cost, the training programme, and his own personal attitudes to piloting, which frankly is not for everyone.

Incidentally, anyone deciding to go for a trial lesson at a local flying club should not turn up dressed in furlined jacket and flying boots. The modern light aeroplane has an enclosed cabin, fully heated and seldom less than clean and comfortable. In short, wear what you would normally wear in a motorcar. The days of wearing heavy, cumbersome clothes in an aircraft's open cockpit have long since gone—except, of course, in light aviation's own grand and exciting vintage world.

At Elstree, membership and joining fees total about £40 and lessons in a four-seater Piper Cherokee, a type popular with flying schools, works out at about £14.50 per hour, plus VAT. Solo flying £12 an hour. Also provided is a 'pilot's bag', with certain navigational reference books and suggestions as to further publications for home study.

Lectures on the subjects required to qualify for the Private Pilot's Licence test cost £1.25 a session.

Although experience is recorded on an hourly basis, flying time is calculated 'chock to chock' or, more familiarly, from the time

Page 35 An International class Kestrel 19 metre wing-span high-performance sailplane—a single-seater with an all-up weight of only 1,040lb

Page 36 (*above*) A Rogallo-type sailwing demonstrates the new sport of hang-gliding over Marlborough Downs; (*below*) ready for the launch—a competitor in the National Parascending Championship

the aircraft leaves its parking place on the apron to the moment the brakes are applied in the same area on return. This gives around 50 minutes of flying time per lesson, a period considered adequate for a student, certainly early in the course when nervous stress and concentration may be more intense than in the latter stages.

Once a pilot has qualified at Elstree, he can by arrangement hire an aircraft at something like £2 an hour less than the going rate for tuition. This is a normal self-drive arrangement with the operator providing the aircraft, fully insured and with fuel included—an arrangement known as a 'wet lease'. This may suit the occasional flier who either cannot or does not wish to buy his own aircraft.

Another range of costs and facilities is provided on the other side of London at Biggin Hill, an acknowledged centre of light aviation and the home of several flying clubs giving instructional courses for the Private Pilot's Licence. Here the Air Touring Club operates the standard 40-hour course, and this table illustraves overall costs:

	£
Club entrance fee	1.00
Club annual subscription	15.00
Hire of club aircraft for 40 hours at £11 an hour	440.00
25 hours (average) instruction at £2 an hour	50.00
Fees for issue of licences (Student Pilot's Licence and Private Pilot's Licence)	7.00
Total (including VAT applicable)	513.00

The entrance and annual subscription fees are payable on joining, but the hourly rates can be met on a pay-as-you-go basis. Note that there is no overall charge for the use of the aircraft, but a two-tier system whereby the instructor's fee is added to the basic hire charge, so that when the student starts to fly solo he merely hires the aircraft at 'wet lease' rates.

A strict hour-by-hour comparison would be unfair in this instance because of the different types of aircraft used. For example, the Piper Cherokee used at Elstree costs more than £12,000 with full avionics and has a larger engine than the Air Touring Club's

c

French-made Rallye Club, which has a price tag of about £7,000. Operating costs, insurance and maintenance charges would further distort any attempt at comparison.

The Air Touring Club was formed at Croydon in 1958 and has built upon its reputation at Biggin Hill with good club facilities, a bar and restaurant being open to members. Regular visits are arranged to the Air Traffic Control Unit at London (Gatwick) so that pilots may be given an appreciation of the controller's problems. The club also arranges flying competitions and 'Breakfast Patrols' to Le Touquet on Sundays. The social side of the club is run by a members' committee, and those who wish to belong to the club without intending to fly are admitted on a half-membership fee. In addition to the PPL course, the Air Touring Club offers tuition in IMC Rating, Night Rating and Aerobatics.

The West Wales Flying Club, based at Swansea Airport, is only seven years old, but has built up a strong membership of 120. The club uses American Cessna aircraft, a type familiar through its high-wing configuration. While flying training courses cover PPL, Night Rating and IMC rating, the club specialises in overseas training flights, operating to the Continent, and as far afield as Spain. Its geographical situation brings the club within easy reach of the Channel Islands and Eire.

Flight simulator facilities—a mock-up of a cockpit fitted with controls and instruments—are available at some clubs, or can be arranged. It is expensive equipment and one would expect to pay additional fees of more than £5 an hour for this specialised form of ground training. Hours spent on a flight simulator do not qualify towards the minimum experience required for a PPL.

Audio/visual aid units, using 35mm film strips synchronised to a standard-type tape cassette, are incorporated in the ground training at the Rogers Flight School at Cranfield, in Bedfordshire. An innovation from the United States, the equipment is marketed by Cessna and the programme, which is based on the use of the single-engined Cessna 150, introduces the pupil to the handling of the aircraft's controls and assists his progress to PPL level by a series of pre-flight film lessons.

Finally, two general notes on the selection of a flying training school. First, personalities play an important part in a student's progress and not every pupil and instructor may get along well with each other all the time. One should not be too hasty in complaint, but if there really is incompatability or a lack of communication, a quiet word with the Chief Flying Instructor may remove any misunderstanding or perhaps bring about a change of instructors. (The instructor might also be relieved!)

Second, if instead of the pay-as-you-go system, you think it preferable to accept the discount sometimes offered on pre-payment of the fee for the whole course—a matter of some £500—weigh up the saving against your commitment to finishing the course if you are to get your money's worth, and make sure there is a clear provision for a refund should you decide for any reason not to complete the course.

NAVIGATIONAL STUDIES

For the pupil who wishes to go more deeply into the subject of air navigation, either before or after qualifying for his Private Pilot's Licence, the City of London Polytechnic's School of Navigation opens the way. The school teaches marine and air navigation to full-time students to degree level equivalent, but there are also evening classes in air navigation.

The syllabus has been arranged with the student and the less experienced pilot in mind, and although the course should assist pilots preparing for an instrument rating, it is not intended as a preparation for any particular examination. The outline programme reproduced here, by permission of the London School of Navigation, may be revised according to the particular interest and requirements of those attending. The course consists of lectures and classroom practical work, and students have the opportunity to use the Flight Procedures trainer.

London School of Navigation: Outline Programme

Pre-Flight Planning:
Chart selection. Measurement of tracks and distances. Triangle of

velocities. Calculation of drift and groundspeed. Computer calculations, time, fuel, etc.

Met Reports and Forecasts:
Pre-flight weather considerations. Reports and forecasts. Station circle.

Direction on the Earth:
Earth's magnetic field. Compasses. Variation and deviation. Directional gyro.

Map Reading:
Principles. Pinpoints. Line features. Transit bearings.

Dead Reckoning:
The 1 in 60 rule. Drift lines. G/S checks. Calculation of headings and ETAs.

Position Lines and Fixes:
Types, transferring, construction of fix.

Aircraft Pressure Instruments:
Altimeter errors. Pressure settings. ASI errors. TAS calculations.

Radio Aids:
ADF. Ground DF. VOR.

Airways Procedures:
Joining airways. Flight plan. Airways flying.

Terminal Procedures:
Patterns. Holding. Signals square.

Maps and Charts:
Form of the earth. Projections. Scale. Uses and limitations.

Meteorology:
Airmasses. Fronts. Pressure systems. Associated weather.

Procedures Trainer/Morse Practical/Plotting:
Divide into groups for each activity. Trainer available from 6 pm.

Enrolment forms may be obtained from The Registrar, School of Navigation, City of London Polytechnic, 100 Minories, London EC3N 1JY. The fee for the course is £4.50.

Apart from this course, the school has a very flexible approach towards helping individual students and will arrange lessons and tasks for the student who wishes to attend at specific times during the day, or even lunchtime periods.

RAF SCHOLARSHIP SCHEMES

Under the provisions of the Royal Air Force Flying Award scheme it is possible for a young man to be taught to fly at government expense, without any strings attached and at only a nominal charge. The scheme offers annually up to 350 flying scholarships to youngsters interested in flying and who are members of the Air Training Corps or the Combined Cadet Force/RAF section. The acceptance of an award entails absolutely no obligation to join the RAF.

An applicant must be at least sixteen years of age, although flying training cannot begin before the seventeenth birthday; must still be at school; have GCE 'O' level or equivalent in English language and two other acceptable subjects (or a headmaster's statement that the applicant is expected to pass these standards), and must be a British citizen.

Four-week training courses at Easter and during the summer are held at selected civilian flying clubs. The courses give 30 hours tuition, so that to gain his Private Pilot's Licence the student would have to fly a further five hours at his own expense, although local authorities may consider a grant to cover the additional outlay, dependent on parents' income. About 75 per cent of flying scholarship students go on to civil PPL standard.

The period of 30 hours tuition is a hangover from the days when the then Board of Trade-approved course specified that time as the minimum qualification for the PPL. It has yet to be amended, but revision may soon bring RAF-sponsored courses into line with the current Civil Aviation Authority requirements.

Although civilian clubs and not military establishments provide the tuition, RAF officers check out all the instructors involved at least once a year, and keep an eye on the standard of student flying. Cadets must, however, undergo pre-assessment tests at the Officer and Aircrew Selection Centre, RAF Biggin Hill. These medical, aptitude tests and interviews last three days, and while the medical examination is to normal service standards, it does not have to meet the more stringent requirements of aircrew members serving in the RAF.

Equally, the passing of the Biggin Hill tests in no way guarantees the cadet a subsequent service or aircrew entry into the RAF. About 40 per cent of all applicants pass these tests to qualify for the Headquarters Air Cadets Flying Scholarship Competition, where a 'paper' board is held. Main points examined are the applicant's Air Training Corps record and, not unnaturally, recruit potential. Successful cadets for Easter courses are notified in February ánd those for summer courses in May. Successful candidates under eighteen years of age must obtain their parents' or guardians' consent and agree to pay £14 towards the cost of board and accommodation. The RAF pay all other expenses, including travel to the flight centre.

Another RAF-sponsored scheme offers university students the opportunity to learn to fly. This scheme is administered by the Headquarters University Air Squadrons, RAF Cranwell. Until 1973 the triangular navigation tests—involving landing away at other airfields and a necessary qualification for the civil PPL— was not part of the University Air Squadrons' navigation syllabus and few university students were going on to take a Private Pilot's Licence. The syllabus has now been amended to include this qualification, so that all the University Air Squadron flying student has to do is to take the Aviation Law ground examination and convert his University Air Squadron licence to a suitable civil type.

If a young man is attending a school that has no Air Training Corps or Combined Cadet Force/RAF section, or is at a university or college without a University Air Squadron, all is not yet lost. To cover this eventuality, the RAF sets aside a further one hundred places each year under the Special Flying Award Scheme, the object being to attract recruits to practical flying and to a career in the RAF. Once again there is no obligation to enlist, and conditions of eligibility are much the same as for the other two schemes.

The specially devised syllabus for the 30-hour course held over 28 consecutive days allows time for intensive ground study of subjects such as navigation and air law. For the student, the course is effectively a gift of over £400 at current (1974) rates of tuition. In addition, food and accommodation costs makes the whole grant a worthwhile £500.

Unfortunately, these government-financed schemes are aimed at young men only. There are no provisions to attract girls, simply because the RAF does not employ women pilots on service duty. An opportunity for girls to receive free flying training is, however, afforded by the Air Wing of the Girls' Venture Corps, formerly known as the Women's Junior Air Corps. Units in various parts of the country organise instructional courses in different aspects of aviation, such as navigation, theory of flight and air-traffic control and, for the fortunate few, there are a number of flying scholarships to be won which entitle the holder to free tuition at a flying school in order to qualify for a Private Pilot's Licence.

CHAPTER THREE

RULES OF THE AIR

Assuming that the coveted licence has now been gained, let us examine some of the responsibilities as well as the privileges of the private pilot in the United Kingdom.

First, he or she is not permitted to take on aerial work or ply for hire. In short, one may not receive any remuneration from one's flying. A private pilot may, however, by arrangement with a flying club, give instruction to fellow members, and the regulation does not preclude him from flying an aircraft for his own business purposes.

Second, if a pilot has a fancy to overfly a sporting event—say to watch his local football team do battle or cheer his fancy home in the Derby at Epsom—he will be breaking the law. Rules of the Air and Air Traffic Control Regulations (1970) specify that an aircraft must not fly over or within 3,000ft of any assembly in the open air of more than 1,000 people assembled for the purpose of witnessing, or participating in any organised event, unless given permission in writing both from the organisers and the Civil Aviation Authority. For the record, the average First Division gate is about 30,000, and the Derby crowd at Epsom usually numbers half a million.

Third, an aircraft must not fly closer than 500ft to any person, vessel, vehicle, or structure, unless of course the pilot is taking-off or landing. The 500ft minimum is raised to 1,500ft over cities, towns and other urban areas but it is also the responsibility of the pilot to fly high enough above this level to ensure landing on open ground in the event of engine failure. Points two and three can, of course, be ignored in order to save a life.

The fly-on-the-right principle prevails. If two aircraft are approaching head-on, both machines must alter course to the right. When on converging courses, the rule is to give way to the aircraft on one's right. An aircraft being overtaken has the right of way, and the overtaking machine must pass well to the left. Similarly, an aircraft flying within sight of the ground and following a road, railway, canal or coastline, must keep these landmarks on its left—in short, drive on the right.

Now to the subject of airspace over the United Kingdom. To the uninstructed, the controlled areas, viewed from the horizontal, resemble an uneven layer cake. Various heights at which aircraft must fly are specified for different parts of the country. Viewed vertically, the areas controlled, for various reasons, take on irregular shapes, but maps and other information leave the pilot in no doubt as to his freedom and his limits.

Much the same as the motorist, the private pilot soon establishes his own favourite jaunts. Cruising at 100 knots, Britain's coast is within an hour's flying from almost anywhere in the country, and a summer evening's trip to the Isle of Wight is quite simple from most aerodromes to the south of London. The English Channel, that time-honoured challenge to airmen from the days of the earliest aeronauts, is becoming increasingly popular as an airbridge to French and Belgian seaside resorts. The very sound of Le Touquet or Ostend for the day stirs many a weekend flier and flying abroad, whether for the day or for a longer holiday far to the south, now presents few problems and many attractions.

Formalities are kept to a minimum. All the pilot has to do is present himself and passengers at a Customs airport, such as Southend in Essex or Ashford in Kent, for the usual personal clearance. The rule is that before an aircraft can leave the country an export licence must be obtained but this requirement is waived if the aircraft is leaving only temporarily. In this case all the pilot has to do is present a Customs carnet or Customs Form XS 29 on which a written undertaking is given that the aircraft will

return to the United Kingdom within a month. The same procedure operates on the return trip. Should a pilot land other
that at a Customs airfield, the authorities must be notified, and
Customs consent must be obtained before the aircraft unloads or
leaves the area.

A refund of Customs duty is granted on the fuel in the tanks
of private aircraft going abroad from a Customs airfield. The
qualification for this concession is that the pilot later provides
proof of being cleared by foreign Customs. This is not a limited
concession and if a pilot chooses to fly daily or weekly across the
Channel, all fuel in his tanks before the outward journey would
qualify for the duty refund.

There are certain local restrictions for a private aircraft in most
countries and as these are subject to change at short notice the
private pilot would be wise to contact the Facilitation Department
of the British Light Aviation Centre when planning any overseas
trip.

THE POPULAR FLYING ASSOCIATION

Most leisure activities involving a vehicle, whether it be sailing,
motoring or aircraft, usually attract a small band of enthusiasts
who are not content to use standard production models, but
prefer to build a vehicle for themselves.

The Popular Flying Association, now based at Shoreham Airport, in Sussex, represents, along with other activities, this type of
private aviator. A non-profitmaking organisation open to anyone
interested in aviation, its prime aims are to get the private pilot
into the air as cheaply as possible and to promote the design and
construction of ultra-light aircraft.

The association was formed in 1946 as the Ultra Light Aircraft
Association, at a time when immediate post-war restrictions were
far from conducive to private flying. Today, the PFA works
alongside the Civil Aviation Authority and the Air Registration
Board in issuing 'Permits To Fly'. The association's engineer
executive examines new aircraft designs from both commercial
and private members and advises or approves modifications. List
of stockists of building materials are available, in addition to

lists of plans from British builders. A national network of PFA inspectors has been set up to assist builders and to sign out their work in preparation for a 'Permit To Fly' application. 'Permit To Fly' certificates may be granted to ultra-light aircraft, or aircraft under a certain weight, where the owner is allowed to build or maintain his own machine, outside the formal Certificate of Airworthiness conditions demanded by the CAA.

The association recommends the home builder without design experience to use an existing approved design, that is one for which stress figures have already been checked out by either the CAA, the PFA in Britain, or a similar body abroad. The construction itself must be registered and arrangements made for an aircraft inspector to check the building progress throughout. Should the amateur builder aim for the ultimate and not only build but also design his own aircraft, the PFA will advise him as to structural requirements. It also recommends that CAA-approved engines should be used wherever possible and some home builders convert engine units for the purpose. Single ignition is not acceptable in aircraft other than powered gliders, which are defined by a certain power-off sink rate.

The PFA was the originator of the co-ownership group system of aircraft operation, and a number of these groups are now actively engaged in aircraft construction. The association is also concerned with the preservation of vintage aircraft, maintaining a record of types and information as to the whereabouts of precious spares for vintage aeroplanes.

CHAPTER FOUR

THE HELICOPTER

Few methods of transport today can compare with the helicopter for versatility and manoeuvrability, but while the high costs involved can usually be justified for business purposes or for specialist operations, for pure leisure pursuits the helicopter is expensive from every aspect, be it tuition, hiring or buying.

Many helicopter training schools use the American-built Bell 47 series, that familiar wasp-like model which appears to consist of a large perspex bubble perched at one end of a lattice-type steel frame. Fragile though it looks, this machine might cost £26,000 plus avionics, or perhaps three times the price of a light aircraft with a similar three-seater configuration. The helicopter has to meet the Certificate of Airworthiness requirements laid down for the type and servicing costs are much higher than for fixed-wing aircraft. Fuel consumption is probably twice the rate of its fixed-wing counterpart of similar capacity. The Bell 47G-5A, for example, consumes something like twelve gallons an hour at its cruising speed of 74 knots. (Picture, p 18.)

Aviation insurers quote higher premiums for helicopters, even allowing for the extra value of the machine, and the insured is usually also required to pay a substantial portion of any claim for damage. Helicopter instructors, too, being very highly trained and often ex-Service pilots, command considerably higher salaries than their light aircraft counterparts. Saddled with these fixed costs, the school adds in administration charges and profit margins and comes up with an overall tuition scale of something like £42 an hour plus VAT.

The Civil Aviation Authority lays down a minimum period of tuition in line with fixed-wing requirements, that is 40 hours, or

35 on approved courses, while holding a Student Pilot's licence. The student therefore can expect to pay in the region of £1,800–£2,000 for his Private Pilot's Licence (Helicopters/Gyroplanes). The £42 fee is an hourly charge on a pay-as-you-go basis, but payment in advance may bring an effective £2 hourly reduction. A course is usually inclusive, covering ground instruction and lectures on airmanship, meteorology and traffic control, aviation law, and navigation, a syllabus similar to a fixed-wing pilot's course. The principles of helicopter flight are, of course, of a specialised nature and the student receives comprehensive lectures on this subject. Schools prefer students to undertake an intensive course spread over two or three months, but many choose to spread the tuition over a longer period. Once qualified, the student is able to take advanced instruction in all aspects of helicopter operation.

HELICOPTER LICENCE REQUIREMENTS

The requirements of the Private Pilot's Licence (Helicopters Gyroplanes) in respect of examinations, licence fees and conditions and other tests are much the same as those for Private Pilot's Licence (Aeroplanes). Again the minimum age for the granting of the licence is 17 years, and the applicant must be medically examined by a doctor authorised by the CAA. An applicant already holding an aeroplane licence is permitted to take the helicopter course with five fewer qualifying hours than would otherwise be required.

The breakdown of the 35- and 40-hour courses specifies a flying experience of at least 12 hours dual instruction and at least 10 hours as pilot-in-command of a helicopter, including a cross-country flight of 25 miles and a landing at a distant landing ground. There is no general flying test for a helicopter pilot's licence but an examination must be carried out by an authorised rating examiner on the type of helicopter which the applicant has been learning to fly. Generally, a qualified pilot converting to a different machine will need five hours experience and an endorsement for the new type.

The authorised flight examiner will conduct the test, by day, through a series of drills with which the student should be well acquainted by the end of his 35- or 40-hour course. The CAA lays down the following pattern for the examination:

1. Pre-flight inspection.
2. Starting procedure; running up.
3. Taxying.
4. Take-off, hovering, and landing into wind.
5. Flying a square pattern with constant heading at speeds not exceeding 25 knots.
6. Take-off, turn 360 degrees each way in hovering flight; crosswind landing within the limitations of the type of helicopter.
7. Straight and level flight at predetermined power settings and airspeed.
8. Climbing and descending turns.
9. Steep turns at constant altitude and airspeed.
10. Entry into auto-rotation; overshoot procedure.
11. A landing in simulated auto-rotation in a predetermined position.
12. In servo-controlled aircraft, an approach and landing using the supplementary system.
13. Recognition and correction of 'over-pitching'.
14. Limited power take-off and landing.
15. Action in the event of fire in the air.
16. Flight into, and out of, a restricted landing area.
17. Shut-down procedure.

Subjects for the written ground examinations are (i) Aviation Law, Flight Rules and Procedures, (ii) Navigation and Meteorology; oral examinations are (iii) Rotorcraft (General) and (iv) Airframes and Engines (Specific Type).

Auto-rotation, as referred to in the test list, is probably the only term that needs amplification. The sequence itself is a very reassuring one for it answers the common question of 'What happens if the motor stops?'. To the novice it would appear inevitable that, once having lost its lift qualities and without wings on which to glide, the machine must plummet earthwards. But to a helicopter the great swirling rotor blades are the wings, and even though the engine ceases to provide the power for upward pull, the air flow over the rotor blades continues to give considerable lift from below. The rotor blades continue to rotate, driven automatically by the air pressure just as a sycamore leaf will gently gyrate earthward. The pilot is thus enabled to select a glide angle according to landing facilities, with the machine descending at some 600–800ft a minute—a gentle sink rate which is much slower than the 20ft per second rate of a 12-stone parachutist.

When the student pilot first qualifies for his full helicopter licence he is not permitted to fly out of sight of the ground or water, or by sole reference to instruments, until he has passed a course of blind-flying instruction. Usually five hours with a qualified instructor will enable him to acquire an instrument endorsement, after satisfying a helicopter instructor that he is competent in instrument flying. Unlike normal fixed-wing requirements which specify instrument ratings, there is no equivalent in helicopter flying and an Instrument Rating on aeroplanes is not applicable to helicopters.

There is, however, a formal Night Rating in helicopters. The conditions are that the applicant must have at least 50 hours experience, with 25 hours as pilot-in-command; 5 hours instrument tuition, of which half may be served on a simulated flight trainer; and at least five hours qualified instruction. He must also produce evidence of a minimum of five hours as pilot-in-command by night, executing certain landing and take-off drills, hovering, climbing and other manoeuvres; and certification by a qualified helicopter instructor.

FLIGHT RESTRICTIONS

Once qualified, the helicopter pilot has considerable freedom of operation for his highly versatile machine. He does not need an airfield on which to land and there is no ruling to prevent him from operating from private property within the limitations of safety. Aviation law requires that a helicopter must not fly below such a height as would enable it to alight without risk to persons, animals or property in the event of a power unit failure. However, a helicopter is not permitted, without official permission, to fly over a large built-up area, as presented by a city, town or settlement, below a height of 1,500ft above the highest fixed object within 2,000ft of the helicopter.

Certain areas of Central London are also forbidden to the helicopter. Special lanes over London are set out for the exclusive use of helicopters, these lanes being numbered and prefixed by the letter 'H' on certain controlled flying-space maps. In Central London it is not by chance that the river Thames is the main thoroughfare for helicopter traffic. The water surface and the ground below high-tide level present a natural emergency landing area in the event of a forced touchdown.

Many helicopters are equipped with special 'popout' canvas floats which are attached to the long landing skids. These can be inflated immediately by the pilot should he need to come down on water. The inflated floats are designed to support the machine and full load for an indefinite period.

Wherever possible, helicopter lanes are routed over water and open spaces and the usual procedure for a helicopter intending to enter the Central London area is to approach the outward end of the appropriate lane, report and receive control permission to continue the inward flight. For example, Helicopter One's terminal point is at Woking, in Surrey, more than 20 nautical miles out, but less than ten minutes flying time from Battersea Heliport, on the south bank of the Thames. The route from Woking follows the railway line to Sandown Park racecourse, skirts Hampton Court's open spaces, travels across Richmond Park to Barnes Reservoir before turning east to follow the line of the Thames to Battersea. Given the visibility, the pilot will usually fly at around 800ft and, unlike his fixed-wing contemporary, he can always slow down to walking pace if faced with navigational problems.

Page 53 Hot-air balloons under inflation at the start of a rally. The liquid propane gas burner sends a 6ft flame into the envelope in intermittent burns

Page 54 Fully inflated and on the point of lifting off, the 84,000cu ft capacity hot-air balloon *Jules Verne* belonging to a South of England syndicate

CHAPTER FIVE

FINANCE FOR FLYING

INSURANCE ANGLES

Aviation insurance is a specialised field. Premiums do not follow a pre-determined scale, but are based on a wide number of factors and actuarial experience. For the private pilot the aim is adequate cover. Leave a loophole and you are apt to fly straight through it into a situation in which it might prove difficult to obtain redress or, worse, put yourself in a costly position if sued by a third party. Insurance is not cheap, but neither are aeroplanes, and it is an item which adds considerably to the fixed costs of running an aircraft.

Strictly speaking, there is no legal requirement for light aircraft operating in the United Kingdom to be covered for third-party risks, as is required of anyone taking a motorcar on to the public highway. However, no responsible pilot would put himself at financial risk by taking-off uninsured in a machine that might have cost him ten, twenty or thirty thousand pounds. If he crashes, even though unharmed himself, he could bring down power cables to an intensive farm unit or industrial plant and be besieged on all sides with third-party claims and probably, even more staggering, consequential loss claims for damage to livestock in the farm unit or output from the industrial plant. Or, should he choose to take a friend for a Sunday morning flight and that friend be returned safe and sound—before slipping off the wing on alighting—he may find himself faced with a compensation claim for a broken leg and loss of earnings. Friends are apt to become very formal where insurance liability is at stake . . . and possibly less than friendly if you have been negligent in taking care of accident risks.

The first item to be considered is hull insurance, that is the aircraft as it stands, but make sure that the cover includes avionics, that is radio, navigational aids, and other accessories. Such highly sophisticated technical aids could, if you have fallen under the spell of a zealous avionics salesman, easily be worth more than half the original price of the aircraft. And experience has taught the insurance companies to take this point into account.

Hull insurance, then, is expressed as a percentage of the total value of the aircraft. This can start as low as 2 per cent, but is based by the broker on several factors: (i) where the aircraft will be hangared; (ii) whether it will be used for private and/or business purposes, and how many hours it is expected to fly each year; (iii) the number of qualified pilots who are likely to use the machine; (iv) the experience of the pilots concerned and their flying record from an accident-free aspect; (v) the age of the pilot or pilots; (vi) the type of aircraft; (vii) the insured value of the aircraft.   Where the aircraft is to be kept will give the broker an indication of the risk of scraping other aircraft while parked, and whether the machine is to be operated from a private airstrip which is almost certainly to be on grassland—not that the insurance world have anything against grass. Although there is no general added risk to business uses, a pilot-farmer who plans low-level crop-spraying may find himself paying a little more for the privilege of his labour-saving hedgehopping.

The type of aircraft and its value, and whether it is single- or twin-engined will obviously affect the calculations. To the aviation broker, two engines must logically be safer than one and so will help to keep a premium down. Generally, the higher the cover the lower the premium when expressed as a rate. Hull insurance then covers physical damage to your aircraft whether in flight or on the ground. Repairs today are just as expensive for aircraft as they are for motorcars, and even a minor dent caused when taxying may result in a £100 bill from the workshops. While the premium may be assessed at say 2–3 per cent of total value, it usually takes into account the standard practice of deductibles.

Deductibles, or excess clauses, are similar to those in motor insurance but with the important difference that the deductible is based on the value of the aircraft and not expressed as a percentage of an actual claim. In this instance the owner accepts the first impact of claim, according to his agreement on deductibles. To many pilots, and particularly to flying clubs with their own engineering workshops, this is a very satisfactory arrangement. For a comparatively small accident repair job, costing say £100–£200, the club will put the job straight into the workshops, so avoiding paperwork, administration and the time-lag of an insurance claim acceptance before work may start on the damaged aircraft.

A club aircraft, or perhaps the individual pilot who leases his machine to a club, could face a considerable loss of earnings if it were to be grounded for any length of time, especially during fine weather at weekends. The motor no-claims bonus principle also operates with aviation, but instead of a progressive increase in the no-claims bonus the saving comes through a decrease in the annual rate of insurance. A no-claims record is not usually transferable, and many owners prefer to forgo a no-claims policy clause for a slightly lower premium and the freedom to change their insurer, if they wish, without loss of any bonus.

Third-party liability is quoted on a flat premium, usually working out at something like £50 for £50,000 cover, although most insurers would advise £75 for £100,000 as being more realistic today. While at first sight this may seem more than adequate to meet a third-party claim for damages directly caused by you, it could fall woefully short in particularly unfortunate circumstances. Even while on the ground, a light aeroplane could conceivably swing hard into a £750,000 executive jet parked on the apron and, quite apart from a claim for the cost of repairs, there could well be an additional claim for the consequential loss of the executive jet's earnings while undergoing repair.

Moving on from the aircraft and third-party interests, we then come to passenger liability, cover for which is based on the passenger seating capacity of the aircraft and excludes the pilot. This liability rate is assessed at £35 for each passenger seat and assures a £25,000 cover for each person. Personal accident and ordinary life policies do not normally cover private flying activities and additional cover will need to be negotiated, either as an endorsement to the existing policy or as a separate contract. Any additional premium required would then be assessed on the pilot's experience or lack of it, his private or commercial involvement, the number of hours likely to be flown annually, single- or twin-engine rating and, to lesser extent, the age of the applicant. Anyone flying as a passenger in an aircraft other than one operated by a licensed carrier should also notify his life or accident insurer, and though this may not necessarily involve an extra premium, it would be noted on his policy.

As a broad rule of thumb, and bearing in mind that as cover rises the premium expressed as a rate comes down, it may be said that if your modest second-hand aircraft is worth £3,000 or £4,000, the owner can expect to pay 4–5 per cent of its value for overall annual insurance, whereas if the price-tag is £20,000, 2½ per cent is more likely to be the rate. If you have any doubts about the company suggested to you by a sales agent or club, it is both sensible and simple to contact Lloyd's insurance brokers, who will put you in touch with a reliable company for a first or second opinion without any obligation to accept a proposal. Bear in mind, too, that arrangements can now be made for premiums to be paid on a quarterly basis if desired, and be sure to inform your broker or insurance company of any accident, however small. And never order work to be carried out under an anticipated claim without the written authority of the broker.

LOANS AND HIRE-PURCHASE AGREEMENTS

In recent years aviation finance has become a fast-expanding business and there now exists a wide range of facilities for the individual who wishes to borrow the money to buy an aircraft of his own choice, rather than be tied to leasing conditions which might restrict him to a particular type of machine or geographical base. Aviation finance can be obtained for the purchase of all types of aircraft, from gliders and club machines to executive jets and helicopters, both new and used, subject, of course, to the usual creditworthiness enquiries or security arrangements that are a necessary part of the business of loaning money. On a more modest scale it can also be obtained, in the form of personal loans, to defray the costs of a Private Pilot's Licence and other courses, to enable the private owner to improve the operating efficiency of his aircraft by installing radio and other navigational aids, and even to help meet the expensive bills for major overhauls and replacement engines that every private owner must eventually face. Credit cards, too, offer a short-term form of financial borrowing, Barclaycards, for example, being acceptable at some airports in payment of aircraft hire, flying tuition and fuel.

Borrowing money for the purchase of an aircraft, whether it be in the form of a hire-purchase agreement or, more advisedly, by means of an aircraft mortgage deed, can however be rather more expensive than some would-be borrowers are apt to realise.

Interest rates—which are quite simply the cost of borrowing money—fluctuate according to conditions prevailing in money markets and economic factors, but the all-important figure when examining the terms of repayments is the true rate of interest, not the annual percentage rate. In hire-purchase deals the interest is based on the initial amount borrowed and is not adjusted downward as the principal, or sum advanced, decreases. The true rate of interest, then, on a loan spread over a 12-month repayment period may be arrived at by doubling the interest rate stated and taking away one. Thus an apparent 15 per cent loan charge would become 29 per cent expressed as a true rate. Repayment periods beyond a year work out at an even higher rate.

A borrower of £1,000 quoted an interest rate of, say, 15 per cent might consider this quite acceptable without appreciating that this was to be the *annual* rate of interest. Being more keen on buying the item in question than on reading the contract, he might overlook the point that the £1,000 loan spread over two years will really cost him considerably more than £300, for he will still be paying interest on £1,000 when actually owing only perhaps the final instalment. So the point that must be stressed is that the interest charge is calculated on an annual basis through-out the period of the loan. After all, at the other end of the finance business, there is someone quite justifiably expecting his invested capital to yield income on an annual basis and finance houses must pay the going rate for commercial borrowing. The obvious, sensible approach, then, to raising money is, first, to seek out a reputable company; second, to make sure that its rates are acceptable and the repayments within your means and, third, to see that there are no small-print clauses. Establishing the first point should rule out the third.

With all types of hire-purchase finance the amount of the initial deposit and the repayment period will depend upon the credit rating of the customer. While deposits are normally between 20 and 25 per cent of purchase price and repayment of the balance spread over approximately 36 months, there may be situations in which the finance company requires a 50 per cent deposit and a two-year repayment schedule. Conversely, other proposals might call for a much smaller deposit and allow a longer period of repayment.

There are a number of finance companies with specialist avia-tion divisions ready to offer loans tailored to fit individual needs. Large groups, such as Slater Walker Finance Corporation, a sub-sidiary of the Slater Walker Group, or the National Westminster Group, and the Forward Group, an offshoot of Midland Bank, all offer such a service and while much of their activities embrace commercial aircraft right up to large jet airliners, they are usually ready to examine a proposition on a small private aircraft.

The arrival of value added tax in April 1973 changed the appeal of hire purchase to the aviation borrower. VAT is now levied on hire-purchase charges for all aircraft where the total hire purchase price is in excess of £2,000. As this total includes interest payments, it applies to the vast majority of aircraft transactions. For limited companies which are registered VAT traders the addition of VAT on the hire-purchase charges will not, of course, have any real effect since the tax will be passed on.

For this reason among others, individuals, as distinct from

registered companies, seeking financial aid for the purchase of an aircraft would find it more advantageous to take out an aircraft mortgage. The finance cost on a mortgage is simply bank interest and carries two advantages. First, VAT does not have to be applied to interest and, secondly and of special importance, the interest payments are allowable for income-tax relief after the first £35. The interest rates on a mortgage will be similar to those of a hire-purchase agreement.

LEASING AGREEMENTS

Leasing aircraft, a form of contract-hire similar to that offered by car-leasing companies where a rental is agreed and the hirer bears certain operating and fuel costs, may now offer less attraction since VAT has to be added to the leasing rentals. However, for a limited company registered for VAT, leasing still has certain advantages. Cash flow is often an important factor to a company and one of the major benefits of leasing is that it will give an even cash flow to capital asset acquisitions. The rentals are equal throughout the period and the whole of the rentals are chargeable against tax. Tax relief, therefore, is the same throughout the period. Outright purchase, on the other hand, or a hire-purchase arrangement, will usually mean a delay of up to two years before taxation benefits are received and, apart from upsetting the cash flow, there may be occasions when a company is not generating sufficient profits to take full advantage of the tax allowances.

Leasing agreements are usually for periods of three to five years, after which the company may, if it wishes, continue to lease the aircraft for a nominal annual payment. At completion, the resale value is taken into account in giving a refund of rentals to the company, or an alternative is to agree lower rentals over the period of the lease with a final large rental payment calculated on the estimated refund. Leasing rentals or contracts are not normally linked to utilisation of the aircraft and no sliding scale exists to reduce costs as flying hours increase. The lessee, therefore, must pay the full rental whether the aircraft flies one hour or twenty hours a week.

The leasing facilities discussed thus far relate to 'dry' leasing, in which all operating expenses are the responsibility of the lessee. Another type is the 'wet' lease, in which the machine is provided ready to fly, often with fuel included in an agreed rental charge, and possibly geared to a minimum monthly hire fee, plus an hourly rate charge. The 'wet' lease is usually based on a short-term contract covering a year, with a minimum utilisation clause written into the agreement. As the 'wet' lease is really a straightforward deal to provide an aircraft under contract, it is in the operator's interest to include insurance and Certificate of Airworthiness overhauls and maintenance in his package price, and arrange for the proper care and upkeep of his aircraft—usually in his own engineering department.

Contract and leasing schemes are operated by a number of aviation companies and charter firms throughout Great Britain.

CHAPTER SIX

GLIDING AS A SPORT

Such is the autonomy enjoyed by the gliding movement in Britain that the Civil Aviation Authority does not consider it necessary to impose an official licence qualification on gliding enthusiasts—and never has done so. Instead, the British Gliding Association, as the governing body of the sport in the United Kingdom, looks after its own students and sets its own standard of proficiency in the form of a syllabus for the Bronze 'C' examination. Successful candidates may undertake cross-country flights, and the various qualifying levels are explained more fully later.

Membership of a recognised gliding club carries with it associate membership of the BGA, whose funds are derived from a *per capita* levy of all affiliated clubs. There are now 91 civilian gliding clubs in Britain and an additional 15 service clubs, plus 40 through the Air Training Corps/Combined Cadet Force youth movement. Qualified pilot members in association with the BGA now total over 9,000 and recent figures suggested an annual 10 per cent growth, although the 1,000 youngsters qualifying each year through the Air Training Corps/Combined Cadet Force membership may present a higher wastage ratio compared with more mature students in civilian clubs. The minimum age for solo flying is 16, though a 15-year-old may fly a glider if accompanied by an instructor.

There has been a remarkable growth in private ownership of gliders in the last ten years. In 1962 there were in Britain 176 club machines against 145 privately owned gliders. Ten years later there had been a dramatic change, with 387 club machines and 505 private. The term 'privately owned' should be taken to mean 'owned by one or more individuals', as gliding syndicates play a prominent part in the sport.

Two or three sharing members, given congenial personalities, can buy a machine whose cost would be beyond their individual financial resources. Upkeep and, perhaps more important, the labour of maintenance, is also easier with more than one pair of hands. Many syndicates operate quite happily for long periods on verbal agreements spliced with tolerance, but if a more formal arrangement is preferred, members of a prospective syndicate could usefully examine the excellent constitution of the Dante Balloon Group in Appendix F, whose compilers appear to have foreseen most possible eventualities, both likely and unlikely.

Another advantage of a syndicate is that the cost of insurance against damage to the glider or any third-party claim can be spread among the members, including any excess clause payments in policies issued at a lower than standard rate on the understanding that the insurer himself defrays an agreed proportion of any claim. The BGA will put club members in touch with an aviation broker specialising in glider insurance and requires all pilot members to be covered for third-party claims up to £50,000.

BGA regulations do not call for a medical examination for glider pilots but a formal statement of health, similar to that of a motor driving licence application, must be submitted. The questions concern, among other points, freedom from epilepsy, disabling faintness or giddiness; adequate vision (being able to read a car number plate at 25 yards, with or without spectacles, is the usual yardstick); lack of a hand or foot, or any other condition which might impair ability to handle a glider safely. Disqualification is not automatic if an answer is unsatisfactory, but the club committee concerned may require fuller details before permission to fly is given. For a person under 18 years, written parental consent is also necessary.

Where personal accident and life insurance policies have already been taken out, the companies concerned should be notified in writing of any intended gliding activities so that the policies can be suitably endorsed. It is unlikely that any additional premium would be required, bearing in mind the excellent safety record of the sport, but if a pilot should wish to take out extra insurance to cover accidents only while gliding, a special policy would then have to be negotiated.

As with power-driven aircraft, Certificates of Airworthiness are also required for gliders, but the checks in this case are administered by the BGA. Examinations are carried out at specific intervals—annually and, more searchingly, after five and ten years. Gliders marketed by United Kingdom manufacturers are sold with a Certificate of Airworthiness, and imported makes are checked against United Kingdom specification requirements. When a new type is introduced, or an existing model marketed with considerable modifications, the glider has first to be submitted to a board for certification.

The BGA maintains a standard of piloting at club level through appointed instructors. Progression for the novice or student pilot is marked by a series of certificates and badges, and tests have to be successfully passed before the student is permitted to move up to the next grade.

CERTIFICATES OF COMPETENCY

Two-seater gliders with dual controls are used for training at clubs all over Britain and after a few hours dual flying instruction the student will be allowed to fly solo to gain his 'A' certificate and entitlement to wear a badge with a single white bird in silhouette on a blue disc. The 'B' certificate, requiring completion of three solo flights with turns in both directions, brings two birds in silhouette on the badge. Up to this stage the student will have been concerned only with gliding back to earth after having been launched into the air by winch or in tow of a car but to obtain the soaring badge—three white birds this time— more demanding tests are set.

As the term 'soaring' implies, award of the 'C' certificate is conditional upon the student continuing to gain height after the launching-cable has been released and displaying a basic understanding of the soaring technique. The soaring phase must be for a minimum of five minutes, although the instructor will probably suggest that the novice tries to stay up for twice or three times longer than the minimum for good measure.

Experience up to this point will have been limited to straight-forward circuit flying describing a square: a launch into wind, a first crosswind leg, a downwind leg, a second crosswind leg and a final approach to the landing point on the original departure course. The next stage in the glider pilot's progress will come when he is judged ready to attempt his Bronze 'C' examination, the successful accomplishment of which carries with it the privilege of breaking away from circuit flying and embarking upon cross-country journeys—in short, a full gliding licence.

The syllabus laid down by the British Gliding Association for the Bronze 'C' examination falls into five sections:

A. Principles of flight
  1. Elementary understanding of:
     (a) Aerofoils
     (b) Lift and drag
     (c) Forces acting during flight
     (d) Turning
     (e) Stalling
     (f) Spinning
     (g) Loading, placard speeds.
  2. (a) Effects of controls—axes
     (b) Further effects of controls
     (c) Stability.

B. Meteorology
  1. Elementary understanding of:
     (a) Fronts. (Recognition of approach, associated pressure changes, passage)
     (b) Convection
     (c) Cloud formations (lapse rates, condensation levels)
     (d) Weather maps (basic understanding of signs, etc)
     (e) Gliding aspects—wave, hill lift, soaring.

C. Navigation
  1. (a) Map reading—conventional signs
     (b) Appreciation of cross-country flying, effect of wind on track and ground speed
     (c) Field landings.

2. (a) Compass—deviation and variation, turning and acceleration errors
   (b) Vector triangles.
3. (a) Geographical appreciation, ie true and magnetic North Poles, lines of latitude and longitude, northern and southern hemisphere pressure distribution.

D. Instruments

1. Elementary understanding of construction and principles of:
   (a) Air speed indicator
   (b) Altimeter
   (c) Variometer.
2. (a) Errors
   (b) Over/under reading causes and effects.

E. Airmanship/General

1. Full knowledge of Air Law
2. Ground handling—all aspects.

These are written tests. In the air, the student must make two flights each of more than 30 minutes duration, demonstrate his general flying capabilities to an accompanying instructor and carry out an accurately judged 'spot' landing. On achieving the Bronze 'C', the newly qualified pilot adds a small bronze filigree to his 'C' badge. There are further qualifications to be gained in advanced gliding, but these will not concern students at this stage, and will be referred to later.

IN SEARCH OF LIFT

Gliders derive lift, that is upward motion, from three principal types of air currents: hill lift, wave soaring and thermal activity. The first two are basically topographical phenomena and the third meteorological. In the early days of gliding, activities were centred on hill lift and almost every club was sited on a hillside. A good example is the Dunstable home of the London Gliding Club, in an almost perfect semi-circular bowl. The chalk soil here is quick draining and quick warming, making the Downs a good soaring site, and a steep slope to one side of the airfield facing into the prevailing west wind forces wind currents sharply upwards, sometimes to a height of several hundred feet, so providing the lift to enable gliders to remain aloft in conditions not normally conducive to soaring.

Wave or lee soaring takes place to the sheltered side of mountainous regions, the Pennine range, in particular, providing fine opportunities for this type of soaring flight. The wind is forced up from the west by the peaks then, suddenly free of the upward pressure from the unyielding contours of the land, comes crashing down on the far side like a great sea wave. In certain conditions the wind rebounds upwards from the lee of the mountains, enabling gliders to ride the upcurrents to heights above 12,000ft. The higher the wave travels the more its strength is dissipated as the air currents curl away down on the lee side to form a further wave some distance away. The waves move across the country in great ripples, slowly losing strength and height on each rebound.

Other areas have their own particular characteristics. The home base of the Lakes Gliding Club, for example, is at Walney airfield at the southern end of the Duddon Estuary, near Barrow in Furness. The estuary, to the west of the Pennines, is about six miles wide and is surrounded on three sides by hills forming a large bowl. To the west is Black Coombe (1,960ft) and a ridge extending northwards ultimately to Scafell. To the north lie Coniston Fells and to the east the 1,000ft Kirkby Moors. When the wind is blowing onto the hills, an aero-tow to Black Coombe or Kirkby enables hill or ridge lift to be utilised, giving sometimes up to five hours flying time. The large bowl is in the lee of one of the ridges in all but southerly wind directions, so that lee wave conditions frequently prevail. Local pilots have achieved heights of up to 18,000ft using wave lift.

Each club uses its local conditions to best advantage, but it is thermal activity that provides lift of a more general and more extensive nature. Thermals are not peculiar to any one part of the country and these currents, formed by heated air rising from the ground, exist in varying strengths and weaknesses for much of the year. As the rising air is warmed by sun-heated ground, it is obviously a seasonal phenomenon. In winter, thermal soaring is strictly for the birds.

The rising thermals, on reaching the cooler atmosphere between 3,000 and 5,000ft, often condensate to produce cumulus cloud and the glider pilot uses these clouds as markers of thermal activity to help him travel cross-country. The theory is that the pilot joins the thermal to gain height, climbing on the upcurrent in a spiral. When he considers that maximum use has been made of the thermal, he peels off and heads in a straight gliding line for the next available source of thermal lift. Under favourable conditions and making skilful use of a series of thermals, a glider pilot can cover surprisingly long distances and speeds of over 100kph have been recorded in long-distance triangular competitions.

CROSS-COUNTRY FLYING

The stalling speed of gliders varies according to design but the average is about 30–35 knots, that is air speed and not speed over the ground. So that gliders can travel for long distances relying only on air pressure under the wing surfaces to support them, they are designed with flat gliding angles. High-performance gliders have a ratio of 1 to 40, or the ability in favourable conditions to travel 40 miles for every mile of height lost on the downglide. The glide ratio comes down considerably on other types, while gliders used in training have a ratio of something like a 1 to 25.

It is on cross-country flights that the high-ratio design gliders really show their superiority in being able to remain aloft while the pilot searches for an elusive thermal. The thermal base varies in diameter and if it is weak at the lower levels the pilot will need to use every ounce of lift he can find. On the other hand, a fairly strong upcurrent may not persist all the way to the top of the thermal. The higher the column rises, the more it may veer in the strengthening wind direction, possibly on the opposite bearing to which the pilot may wish to go, so that it is sometimes necessary to steer a tortuous course from one source of lift to another in order to make good a desired track.

Obviously, it is not always possible for the glider pilot who runs out of 'lift' to return to his home base and if no alternative airfield is within gliding distance he must look for a suitable field or other open space on which to put down. Lacking a radio transmitter with which to inform his base as to his intended landing place, he will then have to find the nearest telephone and thereafter patiently await the arrival of a retrieval party with trailer, followed by a landborne ride back to his club.

Any pilot flying cross-country will always be on the look-out for potential landing places—ideally, though seasonably, a field of stubble with a nice wide gateway to a road or lane. The stubble treats the underside of the glider gently and the gateway affords access for the retrieval vehicle. A code of conduct has been drawn up by the BGA in conjunction with those most likely to be affected by such forced landings—the National Farmers' Union—and pilots are careful to avoid landing among standing crops or animals and to seek permission before bringing a retrieval vehicle and its trailer onto a field.

Gliding appeals mainly to the individual, and it is sometimes said that those best suited to the sport are of the type for whom team games hold little interest or attraction. Its fascination lies in the personal relationship it involves between the glider pilot and the elements, and in the challenge it presents to seek out and harness natural forces unaided by mechanical means. More akin to sailing, gliding cannot be compared with power flying. Swift reactions are not needed, but, rather, anticipation and gentle movements. Relaxation, rather than fiddling, is the key to successful handling of the controls, so that the mind is thus left free to deal with the business of gliding and soaring without conscious effort.

Once in the air, the glider pilot has no particular privileges over his powered counterpart. He must conform with air regulations as regards height over towns and in open country, which means that, except for take-off and landing, or when on a hillsite, he must keep 500ft away from any object and fly at sufficient height to glide clear of built-up areas. He does, however, have the right of way over all powered aircraft and retains this privilege even while on aero-tow—the aero-tug, of course, sharing the glider's priority.

Page 71 A sky-diver photographed in free fall by a fellow parachutist: the rip-cord handle of the main 'chute can be seen on the parachutist's right-hand side just above the sensitive altimeter

Page 72 (*above*) A fifth man tracks in for a high-level link-up in a demonstration by the Falcons, the RAF's free-fall parachuting team; (*below*) now outmoded, the 'pull-off' method in which the fully deployed canopy plucked the parachutist from the wing of a aircraft, in this picture a Vickers Vimy

LAUNCHING TECHNIQUES

The use of light aircraft for towing gliders higher than is possible by winch-launch or car-tow is growing rapidly and has done much both to lessen the glider pilot's dependence on natural lift conditions and to give more time in the air on days when soaring is precluded. While a winch or car-launch may get a glider up to 800ft, sometimes with little prospect of more than a circuit flight back to base, an aeroplane can easily tow a glider to 2,000ft and in some areas much higher, and so ensure the pilot of a flight of at least 15–20 minutes and a chance to search around for a thermal.

A good aero-tow pilot is usually also a good glider pilot. He does not merely haul the glider up to 2,000ft above the airfield but perhaps two or three miles away if the glider pilot has spotted his potential thermal and indicated his wishes before take-off. The charge for the tug is normally proportional to the height finally reached and it is the glider pilot who eventually pulls the release knob, not the tugman.

The novice will be surprised to find just how quickly he is off the ground—a matter of feet rather than yards. On a winch or car launch, where the glider is pulled by a cable, the glider has to rise steeply to gain height, whereas on aero-tow the glider is held to roughly the same rate of climb as the towing aircraft, so achieving a more gentle ascent. Once released from the towing force, the training glider will probably fly at a speed of 35–45 knots. All is not silent, for the airflow over the external surfaces of the glider produces a pleasant hum, rather like that of the low-powered electric motor of a hair dryer.

Aptitude dictates the rate of progress but the average novice will probably need to spend two weeks on an intensive course to reach solo standards. This may represent an effective ten hours in the air, although experience is measured rather more by the number of launches, and a novice usually needs between 50 and 80 launches to become proficient. If this figure appears rather high, it should be remembered that take-offs and land-

E

ings are probably the most difficult as well as the most important lessons to be learned and that launches are sometimes of only two or three minutes duration.

Less exciting, but nevertheless important, is the instruction given in ground-handling the glider in preparation for both your own and other people's flights. Everyone is expected to take part in the teamwork of winching, cable retrieving and signalling, and this applies equally to those who are taking a course as paying students.

HOLIDAY COURSES

Courses are held from May to October at gliding clubs all over the country, and a complete list is obtainable from the British Gliding Association, 75 Victoria Street, London SW1H oJD. Costs vary according to the type and duration of course, and whether accommodation is included. Some clubs have residential facilities, while others will recommend local inns or farms where separate bookings can be made. In some cases, too, arrangements can be made to park caravans for the duration of the course.

For a novice set on flying solo during his first course two consecutive weeks is recommended, and though most of the more than twenty clubs now offering courses are geared to weekly sessions, they will usually readily accommodate the student who wishes to stay on for a second week. With facilities varying from club to club there can be no set scale of charges. For a 'flying-only' course, that is all on-site and including club fees but not meals or accommodation, weekly rates start at £25 according to season and district. At a club with dormitory accommodation and meals included, costs may not rise above £30, particularly at a club operating only at high season. A few clubs, run on more commercial lines, offer courses from March to October and employ full-time rather than part-time instructors. Costs at these clubs range from £45 to £55 a week.

An intensive course is clearly the better way to reach solo standard in a comparatively short space of time, but weekend or daily tuition is also available. On a more individual basis, the hire of a two-seater glider, inclusive of instruction charges, may be calculated at £3–£4 an hour, although the true charge is based on minutes flown rather than by the hour. Winch launch charges work out at 40–60p, and aero-tows at 8–11p per 100ft, and are not normally included in flying charges. There is a further charge for membership of the club, which may vary from £4–£18 in entrance fees and an annual subscription ranging from £8–£30. Part of these charges are sometimes waived for students attending residential courses. Once proficient at solo flying, a pilot hiring a club glider can reckon on a charge of about £3–£4 an hour.

Gliding sites can be very damp at times, particularly in the early morning, and while a student need take nothing more than hardwearing and warm clothing for flying purposes, a good, strong pair of shoes is a necessity, bearing in mind that much work will be done rigging gliders for flying and in general launch-point duties. In the early stages, the sensation of flying some-times induces a feeling of mild travel sickness, in which case the use of tablets to forestall the condition may be advisable, although the instructor should be consulted well in advance of the flight if one is in any doubt. This condition is largely psychological, quite common, and certainly nothing to be ashamed of.

When a pilot buys his own glider or joins a syndicate he will find a second-hand market starting at a few hundred pounds for an older model, possibly in need of some attention. He is, how-ever, more likely to go upmarket for something more expensive, perhaps running well into four figures. Fortunately, once the purchase is made, there is very little depreciation to the original capital outlay. Inflation is at work here, although the benefit will be less apparent should the glider later be sold in favour of a more expensive model whose owner would also have enjoyed the same built-in inflation factor.

Should a pilot aspire to the ultimate and go in for something like a brand new international class Kestrel 19 then, by the time all the accessories, including radio, are fitted there will be no change from £6,000. (Picture, p 35.) Two VHF frequencies for the use of glider pilots have been allocated by the Post Office.

THE SILVER 'C' AND GOLD 'C'

Unless he is really exceptional, a newcomer to the sport should reckon on two seasons spent in gaining flying experience before aspiring to competition class. With the Bronze 'C' class attained, the next step is the Silver 'C' which about 25 per cent of Britain's pilots are entitled to wear. Tasks here include a five hours flight, a height gain of 1,000 metres (3,281ft) and a cross-country flight of 50 kilometres (32 miles).

To join the really elite one must qualify for the Gold 'C'. This requires a height gain of 3,000 metres (9,850ft), a feat that can be achieved only in certain parts of Britain under favourable conditions, and a distance flight of 300 kilometres (186 miles). A sealed barograph carried in the glider confirms height claims, and if the distance flight is flown over three legs rather than in a straight line then photographic evidence is required. This is usually furnished by a fixed camera on the outside of the glider, with which the pilot photographs an easily identifiable feature. Gold 'C' pilots may subsequently gain further distinctions through a system of 'diamond' ratings, a rarified level of achievement to which only a very small percentage of pilots attain.

HANG-GLIDING

While soaring and gliding may traditionally be considered the flying technique closest to natural flight, the enthusiasts in the rapidly growing sport of hang-gliding claim an even closer association, using a sail-type wing. The sailwing is a delta-shaped flying machine of aerodynamic design with a framework usually of thin-walled alloy tubing over which is stressed specially treated nylon, the whole article weighing about 45lb. The wing span is some 22–24ft. From the underside is suspended an 'A' frame, also of metal, which supports a child's swing-type seat on to which the pilot's thighs are clamped before flight. When dismantled, the sailwing can be folded and carried on the roof rack of a small car. Assembly takes about five minutes. (Picture, p 36.)

Although the apparatus has a stalling speed of only 5mph, a 20mph headwind can present a nasty backward landing at 15mph, so experience in handling is vital. In spite of the hazards, which are sometimes compared with skiing, the sport has mushroomed in Britain. The Hang-Gliding Association was formed in late 1973 and within twelve months had attracted a membership of 2,000, including 20 clubs and syndicates. Membership is not restricted to the United Kingdom and there are already many overseas members. Annual subscription is £2, including £100,000 third-party indemnity. The association stresses the Country Code, organises events and approves certain sail-wing machines.

The sailwing costs between £200–£250 but is available in kit form for self-assembly for about £100. Here it should be emphasised that modifications should never be attempted and that the machine should be flown as the designer meant it. The sailwing is strictly for flying into a wind blowing strongly on to a steep hillside, and nothing more than steep inclines should ever be contemplated. Rock faces, cliffs or bridges are potentially dangerous, as are car-towing launches.

Flying techniques are therefore completely dependent on wind forces curling up a ridge to provide lift. The theory is to traverse back and forth above the ridge and flights of two hours duration have been achieved in Britain by this method. The sailwing is rather a cumbersome attachment in the first few stumbling yards before take-off but, once airborne, steering can be achieved by shifting the body weight left or right to turn in either direction and always away from the hillside.

Constant wind strengths of up to 20mph are quite satisfactory but gusting presents a real danger, particularly at the point of launch when the flier may be caught off balance and tumbled headlong down the slope. The sport can be followed all year round allowing for weather conditions and bearing in mind that 'lift' is always weaker in winter.

Finding a suitable and accessible site is not always easy. Permission must be obtained in advance or there may be local opposition. National Trust land and National Parks look natural areas but, once again, the authorities should be consulted to avoid a wasted journey through flying being prohibited by some obscure local rule or by-law.

As the sailwing is exclusively a solo vehicle and tandem tuition ruled out, instruction has to be gained by first watching an expert in flight, and then by a process of trial and error. Practice is first by small bird-like hops of perhaps 3–6ft at the bottom of a hill with a gradient of about 1 in 3. As the student progresses, he moves back up the hill for higher launches. A crash helmet without peak, overalls and good ankle boots are advised.

A flight should never be attempted, either by novice or expert, alone on a site, for even a minor injury sustained while in a remote area could create a serious situation if the individual were unable to move and there was no help at hand.

CHAPTER SEVEN

LIGHTER THAN AIR

BALLOONING

To those who are unacquainted with the techniques and skills involved, the balloons flown today appear to have made little progress since 1783 when two French paper makers, Jacques Etienne and Joseph Montgolfier, filled a large bag with smoke from a twig and straw fire and demonstrated that a balloon could be carried upwards by heat. The early adventurers were at first under the impression that the smoke, not the heat, provided the lift, being unaware of the fact that air expands when heated, so that a given volume weighs less than the same volume of cold air.

At much the same time, experiments were also being made with hydrogen-filled balloons, tightly sealed in contrast to the hot-air method of a fire suspended below the open mouth of the balloon. Professor J. A. Charles, a Frenchman, had a silk balloon devised to carry hydrogen and accompanied by the builder, N. M. Robert, found the result sufficiently effective to climb to 2,000ft in a half-hour flight. The Italian, Vincenzo Lunardi, is credited with the first hydrogen balloon flight in Britain in 1784.

The first balloon crossing of the English Channel was accomplished the following year in a gas-filled balloon by a Frenchman, Jean Pierre Blanchard. His fellow crewman was an American physician, J. Jeffries, and the pair took-off from Dover, heading for Calais. They made the crossing but had to jettison every piece of detachable equipment, including their own clothing, in order to keep the fast-descending balloon clear of the waves.

The first successful crossing of the Channel by hot-air balloon was made in comparatively recent times, when Donald Piccard and Edward Yost flew from England to France in 1963.

The rip panel and trail rope, still very much a part of the ballooning technique in use today, were introduced at a very early stage. The rip panel, a large detachable section at the crown of the balloon to allow of a quick release of gas or air on landing, was devised in the mid-1880s by an American, John Wise. The trail rope, normally lashed on to the outside of the basket but quickly dangled at low heights to drag along the ground, thus reducing weight to slow the rate of descent or to check forward speed, was introduced by an Englishman, Charles Green. Green also effectively experimented with the use of coal gas as a less expensive alternative to hydrogen and was one of a crew of three who made a 550-mile flight from England to Germany in a hydrogen balloon in 1836, covering the distance in 18 hours—an unprecedented achievement at that time.

HYDROGEN GAS V HOT AIR

Today the widespread use of balloons in weather forecasting and, more particularly, in the United States for high-altitude experiments has, like parachuting, given modern balloonists the side benefits of vast budgets spent on research and development. Even so, hydrogen ballooning is still an expensive hobby—to inflate a 25,000cu ft balloon in Britain costs up to £250 for each flight —which is why the far less costly hot-air ballooning is so popular in the United States and becoming increasingly so in Europe. A hydrogen-filled balloon employs the conventional method of jettisoning sand ballast to ascend and releasing gas to come down, and is capable of achieving longer flights and reaching greater heights than a hot-air balloon, and moreover in complete silence.

For the amateur, then, the hot-air principle is the cheaper and most straightforward means of introduction to the fascinating experience of ballooning. The silence is broken only by the intermittent blast from the burner unit and once the balloon gains height there is little sensation of speed as, carried along by the wind currents, the balloonist in his basket floats peacefully across the changing landscape.

Unlike a hydrogen-filled balloon, the hot-air balloon carries no ballast other than a drag or trail rope so that its pilotage calls for considerable skill and judgment in climbing, levelling off and descending, plus a sound knowledge of meteorology. By studying weather conditions and with the aid of a good meteorological forecast, the expert can work out a flight path based on the anticipated wind currents, which often resemble a layer cake. At various altitudes the wind may be travelling in a slightly different direction and at varying speeds, and by making intelligent use of these variations the pilot is able to exercise some measure of control over his direction of flight.

Ballooning is particularly susceptible to weather conditions and flights would not normally be attempted in areas of thermal activity, in high winds or in the vicinity of cumulo-nimbus cloud masses. Thermals, the joy of the glider pilot, can be a nightmare to the balloonist, giving him unwanted lift to unsuitable altitudes, while clouds are best avoided both for the turbulence they may cause and the pilot's need to keep the ground in view. Balloon ascents are seldom undertaken if the velocity of the surface wind exceeds about 15 knots, while for the inexperienced pilot the recommendation is a wind force of not more than 8 knots. Minimum height restrictions when passing over built-up areas apply equally to balloons as to aircraft and, just as 'power gives way to sail' in the code of the sea, so balloons have right of way over both aircraft and gliders.

LICENCE REQUIREMENTS

To qualify for a Private Pilot's Licence (Free balloons, hot-air) the following requirements of the Civil Aviation Authority have to be fulfilled:

(i) The applicant must pass the Private Pilot's (Aircraft) Licence medical examination and hold a Student Pilot's Licence.

(ii) A minimum of six flights totalling at least 12 hours under instruction.

(iii) A dual test flight with a CAA examiner.

(iv) A solo flight under the supervision of a CAA examiner.

(v) Written examinations must be passed in the following subjects:

    (A) Aviation law and flight rules and procedures.

    (B) Navigation and meteorology.

    (C) Aerostatics.

The fee for the licence is £5, and the medical and written examinations on navigation, meteorology and air law are the same as for a power pilot's qualifications, the additional paper on aerostatics—or more familiarly ballooning—being of a specialist nature.

Instruction in ballooning can be gained as a member of an existing group or syndicate under the supervision of a qualified balloon pilot, possibly by arrangement with a balloon manufacturer, or by taking a course such as that offered by the Europa Balloon Training School, based in Essex and run by qualified and experienced balloon pilots. The Europa school provides full instructional courses for preparing the novice for the pilot's licence and examination. Instruction costs about £16 per person per hour but the school will quote an inclusive fee for the complete course on request and the figures given here should be taken only as a very general guide to the cost of learning to fly a balloon.

BALLOONING SYNDICATES

The alternative to a training school is to join a group or syndicate —the method largely responsible for the present mounting interest in the sport. Assuming a balloon capable of accommodating a three-man (or woman) crew, a comfortable working number for a team would be from six to eight, for while a larger group would reduce flying costs, there would probably be a permanent queue of trainee balloonists growing impatient to get into the basket. Fortunately, there are occasional openings in existing syndicates, for as their numbers swell so groups either purchase a second balloon or divide into smaller units to set up another club, still associated with the 'parent' syndicate.

There is nothing to prevent several would-be balloonists forming their own group. The basic requirement here, finance apart, is, of course, a reasonable assurance of sustained enthusiasm, and to this end it might be best that one member should be elected to train as the first pilot rather than engage an instructor to introduce each student in turn to the basics of ballooning. Not the easiest of tasks, perhaps, to decide who shall get all the fun at first, but it does lead to the group having its own qualified pilot, and therefore its own instructor, that much sooner.

The launching and flying of a balloon are subjects requiring considerable study and a serious, if not dedicated approach. For the pilot, they call for his full concentration at all times to ensure the safety of his crew and to avoid damage to property. Nevertheless, there is still plenty of involvement on the ground for the rest of the group. The more hands to help with launching the better, and while a balloonist chooses his take-off point, he will have less option about the landing field, perhaps ten or twenty miles away according to weather conditions and his own inclination.

For the crew of the retrieve car—an ordinary vehicle adapted for towing the equipment trailer—the task of retrieval is complicated by the need to try to follow the planned flight route along main roads, twisting lanes and cart tracks which bear no relation to compass readings or air currents. Immediately a balloon takes-off, so the retrieve car leaves, attempting to track the balloon visually. Other members often follow, relying upon their own estimates of where the balloon is likely to land. Some members of balloon groups confess to never having seen their balloon actually touch down.

Once a group is firmly established it is likely that it will rapidly develop a social side that will help to keep things going when the balloon is grounded due to weather or repair work. A model example of a successful syndicate is the Dante balloon group, based in Berkshire. Some of the members are British Airways employees, including a VC-10 pilot, while others represent a cross-section of industry and the professions. The group's 65,000cu ft balloon, *Dante*, with its blue-and-white-chequered canopy, is a familiar sight over Berkshire, Hampshire and Wiltshire, and its members have turned the upper part of a large barn, loaned to them by an interested farmer, into a clubhouse that resembles the mess of a fighter squadron—comfortable, informal and not without the usual wit adorning the flight notice boards and wall maps.

A constitution covering all aspects involved in the joint ownership of a balloon, including the members' responsibilities on and off the ground and their financial commitments, has been drawn up by a legally qualified member of the syndicate and has proved so sound that it is being adopted by other groups both in Britain and abroad. By kind permission of the group, their constitution is reproduced for the guidance of other balloonists in Appendix F.

THE BALLOONISTS' CLUB

Dante group members, like most balloonists, belong to the organising body of the sport, the British Balloon & Airship Club. Formed in 1965, the club now has more than 400 members paying an annual subscription of £2.50 and its headquarters are at 75 Victoria Street, London SW1H 0JD. The club advises on insurance, government regulations concerning licensing, and airworthiness requirements and also publishes its own journal, *The Aerostat*, through whose columns contact may be made with other members interested in setting up a syndicate.

Evidence of the increasing international interest in hot-air ballooning was strikingly forthcoming when, in June 1974, the BBAC staged the biggest balloon meet ever to have been held in Europe, at Cirencester Park, in Gloucestershire. Eighty hot-air and ten hydrogen-filled balloons included entries from Afghanistan, Holland, Germany and Eire and competitions were held to select a team of four pilots to represent Great Britain in the Second World Balloon Championship in the United States in 1975.

INSURANCE REQUIREMENTS

As a general guide to the cost of third-party insurance for the average club balloon, the BBAC mentions a £40 premium for £50,000 cover, but considers this a minimum amount which would not necessarily cover claims for a consequential loss. The club gives the illustration of a balloon landing on a greenhouse. The insurance company may meet the claim for repairing the damage done to the structure readily enough, but recompense for a possibly valuable crop contained therein may be another and more difficult matter. The BBAC will no longer accept entries to organised balloon meets if an owner's insurance policy has a 'consequential loss' exclusion clause. Personal life assurance policies do not normally exclude ballooning, but the company concerned should nevertheless be informed of any participation in the sport, and personal accident protection would have to be separately negotiated with the company or through a broker.

No special protective clothing is needed for ballooning other than what the temperature would usually dictate outdoors. Motorcycle-type crash helmets without peak are often favoured but are not mandatory and flying overalls are convenient as well as serving easily to identify balloonists among a crowd of spectators on the ground.

BALLOONING DOS AND DON'TS

The BBAC, well aware of the necessity for good public relations with landowners and farmers, has, like the British Gliding Association, drawn up a Code of Conduct, approved by the National Farmers' Union, for the guidance of members. The code, which is issued as a directive to all balloon pilots, includes the following instructions:

1. Flight Planning
Do not fly unless you are reasonably certain that your flight path will be over country which is suitable for landing. For example, in July and August you should avoid flying over large areas of standing corn in light wind conditions.

2. The Take-Off
Always obtain permission from the landowner before driving onto the field. Check that during the climb immediately downwind

of the take-off site the balloon will not have to fly low over livestock. Brief your crew and any other helpers regarding closing field gates.

3. In the Air

Always fly at such a height that you do not deliberately cause any disturbance to livestock by flying within 500ft of them.

If it appears that livestock have been disturbed for any reason, note the location of the incident and check the cause with the farmer after landing. If you cannot locate the farm, inform the local police station or telephone the local NFU County Secretary (list of NFU County Secretaries is available from the BBAC).

4. The Landing and Retrieve

Select a landing field that should cause the least possible inconvenience to the landowner. Particular care should be taken during the summer months when standing hay or growing cereal crops cover large areas of the countryside. Remember the grave risk of fire at this time.

Before deploying the trail rope, ensure that the ground below and ahead is clear of livestock, overhead power and telephone lines, buildings or other property which could be damaged.

Immediately after landing, take all reasonable steps to discourage on-lookers from coming into the field and trespassing on the farmer's property.

Contact the landowner or farmer as soon as possible after landing. Obtain permission before allowing any vehicles to drive onto the field.

Never make 'tethered' flights or re-flate the balloon in your landing field unless you have obtained permission to do so from the farmer.

If the landowner or farmer cannot be contacted after landing, you must obtain his address or telephone number and contact him as soon as possible afterwards.

Ensure that farm gates are left as you found them.

If damage is caused or the farmer wishes to take further action, exchange names and addresses, including that of your insurers. Ask the farmer to write to you and on all occasions offer to leave your name and address for the farmer in case any unseen damage has taken place.

COSTS OF OWNERSHIP

Qualifications and responsibilities taken care of, the next question is how much is it all going to cost? As an example, a balloon capable of carrying a three- or four-man crew costs around £2,000 and is generally considered to be the best size for club flying. The canopy will have a capacity of 84,000cu ft when inflated, and if this figure seems meaningless, consider that the topmost point of the balloon, properly known as the crown, will be some 80ft above the ground level, just prior to take-off. Hot-air balloons are large, impressive and always colourful.

The only other item involving any appreciable capital outlay is the retrieve trailer, costing around £120, and the towing-bar attachment to a car which a local garage may supply for £15-£20. An altimeter costs £20 or so, and other instruments may be added later. The fuel—liquid propane gas—is carried in cylinders inside the basket and the average balloon will carry sufficient propane gas to give an effective flying range of three to four hours, depending on load. The cost works out at approximately £3-£4 a flying hour.

The general maintenance of a modern hot-air balloon is fairly simple. The canopy can be repaired by sewing, sticking or by adhesive tape, and the manufacturer's handbook describes the correct method of repairing sections. The fuel system works on an elementary principle and, after a short course of training, anyone with some mechanical or engineering experience should be able to service the equipment.

The useful life of a balloon will obviously depend very much on the users. The canopy is the most vulnerable part and calls for careful handling and laying out before inflation. The ground should be examined for sharp stones or other matter liable to cut or damage the fabric, which should never be left lying in strong sunlight for long periods. Given careful treatment and common-sense, the fabric should have a useful life of several years. The power unit, or burner, is constructed of stainless steel and should be returned occasionally for inspection and repair, although day-to-day servicing should be within the capabilities of the balloonist.

A point in favour of ballooning is that an income can be earned by flying at fetes and agricultural shows during the season. These photogenic balloons are great crowd pullers and there is a steady demand for flight demonstrations, both tethered, which the balloon does not like, and in free flights, which it does. Several large companies such as Shell, Esso and Kodak own their own balloons for publicity purposes but there is a considerable market to be tapped to offset running costs and this is quite within the law.

What is not permissible is the use of advertising slogans, and this goes for parachute canopies, too. A balloon may, however, carry the name of a company or owner and balloon manufacturers can supply name-carrying panels to attach to a canopy in a wide range of colours. Any owner or syndicate intending to exploit this source of revenue would be well advised to get into touch with the BBAC before quoting a fee to a commercial organisation and the club will then give advice as to the going rate for such a contract.

BALLOON TYPES AND SIZES

The balloon canopy is made of polyurethane-treated nylon, a material similar to that of a spinnaker sail. Strong and imporous, the material is further strengthened by load-bearing nylon bands, while the gently curving shape of the canopy has the effect of spreading the stress over the largest possible area. There are three main types of hot-air balloons available. The 'O' (12 gores) and 'A' (20 gores) types are constructed on the bulbous gore principle. The smooth spherical 'S' design is generally used for publicity purposes and is less satisfactory for sporting use. Sizes range widely from 31,000cu ft to 140,000cu ft, the smallest costing under £1,000 and, at the other end of the scale, more than twice as much. The following table provides a summary of sizes, capacities and international classifications.

SUMMARY OF BALLOON SIZES

Capacity × 1,000 cu ft Model no	cu metres	International class	No of men	REMARKS
31	890	AX 4	1	One-man balloon
42	1,190	AX 5	1–2	Two-man balloon; also ideal for longer solo flights
56	1,590	AX 6	2–3	Convenient to operate but a bit small for club flying
65	1,840	AX 7	3	
77	2,180	AX 7	3–4	Best sizes for sporting
84	2,380	AX 8	4	club operation
105	2,970	AX 8	6	Six-man balloon, or for long duration or record-breaking flights
140	3,970	AX 9	8	For large loads or exceptional flights

CONSTRUCTIONAL DETAILS

The balloon canopy incorporates a ripping panel and a discharge, or 'dump' valve. The ripping panel is a large section at the top of the canopy secured by Velcro tape or a system of lacing and provides the pilot with the means, on landing, of allowing hot air to escape quickly, so deflating the balloon. A fully inflated balloon is inclined to glide along the ground, giving the occupants of the basket a bumpy ride before coming to rest. The panel rips away part of the crown section when a red line from the basket is pulled and, as a built-in safety factor, the line has to be broken at three tie points before the panel can be opened. Additionally, several feet of the red line has to be pulled down before breaking the ties, so that there is no risk of an accidental or premature pull. The discharge valve, let into the side of the balloon, can be opened or closed in flight and may be used to let out excess air. It is sparingly employed, since hot-air may have to be replaced and fuel costs money.

The lower part of the balloon canopy is attached to the power

F

unit, the burner load frame, by stainless steel wires whose length has been calculated to preclude risk of burn damage to the nylon fabric. A skirt of glassfibre is attached to the mouth of the canopy, which has the double role of protecting the upper fabric and helping to funnel the hot air blasts, particularly during inflation on the ground. As an alternative to glassfibre, increasing use is being made of a flameproof material, Nomex, which is also used as a material for racing drivers' suits.

The high-output propane gas power unit works on the same principle as a Primus burner, incorporating a vaporising coil and a set of four jets through which the vapour passes to form a flame. The burner is controlled by duplicate on/off valves and high-pressure hoses connect the fuel tanks or cylinders with the burner unit. The four cylinders, of steel or aluminium, are strapped in the corners of the basket and the master cylinder has a separate vapour valve which provides the feed for the pilot burner. The stoutly woven basket is constructed from willow and cane and suspended by stainless steel wire ropes that pass in a continuous loop under the basket. A 150ft long trailrope is strapped in a coil outside the basket and, as a safety measure, a fire extinguisher of the dry powder type is usually installed on the inner wall of the basket.

From outside and at the top of the balloon runs the crown line, a rope that assists in stabilising the balloon during inflation. The crown line must never be anchored to the ground but always held by a crew member, who releases the line when his weight alone can no longer hold the inflating balloon down. Well-meaning assistance at the crown line would be misplaced, for the pilot calculates his rate of inflation on the basis of the accepted weight on the crown line which, of course, is out of sight of the pilot once the hot air begins to inflate the envelope.

ASCENTS AND LANDINGS

One of the many virtues of ballooning is that the whole operation is largely self-sufficient. There is no call for a towing aircraft, winch or runway, merely an open field with some shelter upwind

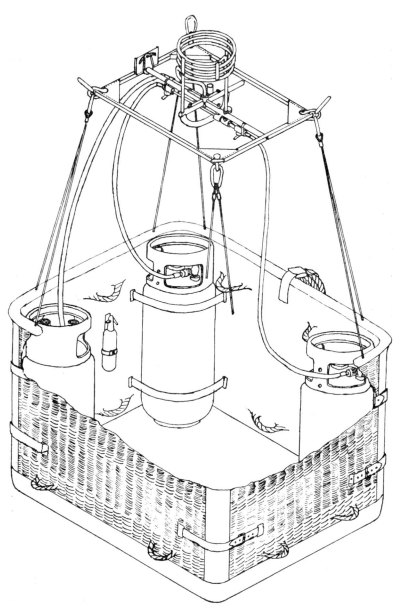

Cutaway drawing of a hot-air balloon basket showing propane gas cylinders and burner assembly

provided by trees or a sharp rise in the ground, and a downwind path free of high obstacles. Balloonists will choose favourite launching spots for several reasons. First, an amenable landowner or farmer, ready to permit the launching from his property; second, knowledge of the immediate terrain and the whereabouts of potential dangers such as pylons, cables and expanses of open water; and third, familiarity with local air space restrictions by the civil or military authorities and, of course, prevailing wind conditions.

A balloon generates almost as much interest on the ground as in the air, and once the balloon envelope is run out from its bag, help will usually be needed to prevent onlookers walking on the fabric, or getting too close to the burner during inflation. Initially, the mouth of the balloon is 'flapped' by two or three helpers to get some air inside the canopy. When the burner is activated by the pilot, who at this stage has both burner unit and basket horizontal to the ground, the mouth is held wide open to admit the short, sharp, noisy burns which send a six-foot flame into the balloon. The pilot gives each intermittent burn a few moments to bubble round inside the canopy, aiming at a steady, overall inflation with air heated to a temperature around 100–120 degrees Fahrenheit. (Picture, p 54.)

The crucial part of the inflation comes when the balloon lift reaches the weight of the one man on the crown line. Plenty of assistance is required to anchor the basket and envelope when the crown line is released and the balloon slowly assumes a vertical position, with the pilot climbing into the basket while still controlling the burner. The experienced pilot can 'feel' the lift of a balloon immediately prior to take-off and, when the atmosphere is still, the height can be controlled to an accuracy of a few inches. In ballooning there is a traditional drill at the point of take-off. Helpers get a 'hands-on' order and bear down on the sides of the basket. This is followed by 'hands-off', when the pilot assesses the amount of lift and adjusts the burner accordingly. A further gentle burn and the balloon, with pilot and passengers, rises to travel at the speed and in the direction of the prevailing wind, once clear of shelter. Bearing in mind that each short blast of hot-air has to rise inside the balloon, a delaying factor has to be allowed for, as continuous burns could overheat as well as being a waste of fuel. In flight, altitude is controlled by adjusting the temperature of the enclosed air by means of the burner.

The pilot can be selective as to his landing point, but only within the confines of his flight path. With the landing code in mind, he will look for a field free of crops, animals or power lines and, hopefully, a reasonably flat, grassy surface. Occasionally he runs out of choice and the occupants of the basket have to hold on to the internal handles and brace themselves for a bumpy touch-down. Hands should never grip the basket edges.

A HOT-AIR DIRIGIBLE

The gas-filled airship, whether it be helium or hydrogen, is far too costly a proposition for other than commercial enterprise or a syndicate of wealthy amateurs, but the hot-air enthusiasts have already tackled the problem of steering their craft and a well-known English balloonist, Don Cameron, has designed and built a prototype hot-air dirigible. Instead of the envelope being bulbous, it is cigar-shaped, 100ft in length and requires 96,000cu ft of hot-air for inflation. The open gondola amidships carries a crew of two, together with the burner unit, and forward motion is provided by a 1,600cc engine mounted at the rear of the gondola and driving a 5ft diameter propeller. Maximum design speed is of the order of 20 knots, and the estimated price of a production model is about £6,000.

CHAPTER EIGHT

PARACHUTING

In the last two decades vast sums have been spent on the development of the parachute for specific tasks, particularly in connection with the American space exploration programme. A side-effect of this investment has been a widespread increase in the popularity of parachuting as a sport in its own right. In the United States sophisticated demands on the parachute, such as the project for the recovery of Apollo spacecraft, have called for considerable research into aerodynamic deceleration, the benefit of which will no doubt be reflected in parachute design and techniques generally.

The development of steerable parachutes has already made possible a remarkable degree of accuracy in landing at a chosen spot and canopies are no longer completely at the mercy of the wind. The vents opposite the cutaway lip of the parachute canopy impart a gliding motion as the parachute moves through the air, much on the same principle as an aircraft wing. Additionally, this lift reduces the landing speed and the expert can alight on his feet with no greater shock than jumping from a garden wall. In suitable conditions, the parachutist can achieve a horizontal speed of more than 12mph, the air flow from the vents resembling the working of a low-power jet engine.

Another advancement has been the incorporation of the sleeve, which holds the canopy and controls the rigging lines during deployment, making it almost impossible for the parachutist to foul lines or canopy during deployment. It also helps to pull the body gradually to an upright position, so reducing the arresting shock. These refinements have produced an exceptional safety record, though it is still a requirement to carry a slightly smaller reserve parachute, and international regulations call on sport parachutists to conform to a strict code of training and practices. Parachute and free-fall displays have done much to bring sport parachuting to public attention and are now a popular and exciting feature of many kinds of outdoor event. Jumping from 10,000ft, the sky-diver, plunging earthward at 120mph, has nearly a minute in which to make coloured smoke trails and join up with his colleagues in various configurations of circles and stars before breaking away to deploy his parachute with a reassuring crack.

SPORT PARACHUTING

Sport parachuting first attained international status in 1951, when five national teams met at Bled, Yugoslavia, in an accuracy jumping competition sponsored by the Federation Aeronautique Internationale. Since 1954, world championships have been held every other year, with more and more countries competing on each occasion.

The controlling body of the sport in Britain is the British Parachute Association, working in conjunction with the Civil Aviation Authority. The association is responsible for the establishment and maintenance of training and safety standards and all parachutists, civilian and military, are required to observe the strict rules laid down for novice and expert alike.

Current membership of the BPA reached a record 5,240 in 1974 and more than 15,000 parachutists have belonged to it at various times since it was formed in 1962. The association organises an annual British national championship and nominates the British teams for international events, including the world championships.

From its London headquarters, the BPA administers full-time and weekend parachuting clubs in many parts of the country, all of which run courses for novices as well as affording experienced members facilities for free-fall practice. At the time of writing there are 4 full-time and 28 weekend civilian clubs and 16 Service clubs in the United Kingdom, in Germany, Cyprus and Hong

Kong. In Scotland, which has its own Scottish Parachute Association, there are three clubs in active operation. A list of clubs open to civilian membership will be found in the reference section at the back of this book.

Membership of the BPA costs £4 for the first year and £3.50 per annum thereafter. Included in the subscription fee is third-party liability insurance cover up to £100,000 for one incident, and the comparative safety of parachuting is further illustrated by the fact that member-companies of the British Insurance Association do not generally suspend life cover for parachutists. It is important, however, that the individual notifies, in writing, his insurer of his intention to follow the sport, which may or may not lead to a slight increase in premium.

A weekend course of parachuting instruction, at the time of writing, costs about £20. This includes BPA subscription, instruction, hire of parachute for the first jump and, sometimes, the use of facilities at the local aero club, which usually provides the aircraft from which the jump is made. Subsequent descents cost approximately £2.50, parachute supplied. The trainee must make a minimum of six descents using a parachute opened automatically by a static line fixed to a strong point within the aircraft. In Britain, parachutists are required to pass a medical examination to ensure that they measure up to the fitness requirements set out in a form which may be obtained from BPA. The medical standard required is simply that of a generally good state of health, limbs sound enough to withstand the shock of landing, a satisfactory mental condition, and freedom from epilepsy or the results of any previous injury which might have a tendency to recur. Spectacles may be worn provided the vision satisfies certain standards. A further medical examination must be carried out every two years, or after any bout of illness or accident. The minimum age for sport parachuting in Britain is sixteen and a parent's or guardian's consent is needed if the candidate is under eighteen years of age.

SKY-DIVING TRAINING

Average training costs and paperwork dealt with, the next question the would-be sky-diver must ask him or herself is: 'Is this for me?'. The sport calls for confidence, both in oneself and one's instructor, as well as a good sense of judgment. Just a few hours orientation at a training centre goes a long way towards instilling a considerable amount of the required confidence and, given average ability, judgment will be found to be simply a matter of experience.

The aerodromes used as centres for sky-diving clubs are often grassy runways, sometimes little more than a landing strip with a clubhouse and a hangar or two to house the club's aircraft. A suspended parachute mock-up usually occupies one corner of the hangar, with a harness suspended from the roof and a 4ft high platform representing the aircraft exit. Weather permitting, the initial ground training will be carried out in the open in the form of lectures ranging from canopy handling to aircraft procedure. Little time is lost in getting the students down to work. Training parachutes are provided by the club but helmets, a motorcycle type without a peak, boots, and light-coloured overalls or jumpsuits are the student's responsibility.

The light-coloured suit is obligatory for the student as it helps the instructor to identify and observe him in the air. Once he has qualified for his General Permit—the equivalent of a parachuting licence—the sky-diver can follow his own choice of colour, bearing in mind that the wearing of bright colours makes the free-falling parachutist more easily discernible both from above and below. Specially designed jumping boots can be bought once the student has decided to pursue the sport and a favoured type is high-laced and has pneumatic soles. Climbing boots, or those with external fittings, are not suitable as the clips may catch in the parachute lines. Goggles may be used, but a one-piece eye mask is preferable if spectacles are normally worn.

An instructional class is usually restricted to eight pupils and the instructor will make sure that every point in his lecture is clearly understood by each member of the class, questioning them individually if necessary. A good sense of balance is vital to the sky-diver and if a student obviously lacks this aptitude there may be a quiet word from the instructor to save the student wasting further time and money. For in sky diving, balance is related to stability, and putting a hand or foot out of place will disturb the air cushion that is created below the sky-diver's plummetting body.

The student first starts to learn the technique lying flat on his stomach in the basic 'spreadeagle' or stable position. In this attitude the terminal velocity, or maximum speed, attainable is approximately 120mph whereas the experienced sky-diver, assuming a head-down delta shape, will reach a speed of over 200mph.

During training and qualifying jumps, the parachute will be opened automatically by means of a static line, usually a line of nylon webbing about 15ft long. One end is connected to the release mechanism on the parachute and the other to a strong anchor within the aircraft. When the trainee parachutist drops from the aircraft, it takes approximately 2·7 seconds for his weight to extend the static line, release and deploy the canopy. The student makes this jump from the aircraft while it is cruising at about 2,500ft over a supervised dropping zone, and in that 2·7 seconds he will fall about 300ft. After completing his initial training course, the holder of a Non-General (ie, provisional) Permit is permitted to complete not more than three parachute descents in a day, and after three full days parachuting must have one complete day of rest free from parachuting.

Much instruction is devoted to the landing procedure. The impact generally is no greater than that experienced in jumping off a 4ft wall, and the student is taught how to minimise its force by various methods, such as flexing the knees or rolling over on impact, and how to keep control of his collapsing parachute.

GROUND TRAINING

A weekend initial training course normally involves about thirteen hours of concentrated instruction and exactly what this comprises is shown in the following minimum ground training programme recommended by the British Parachute Association:

1. Orientation (30 minutes)
    (a) Documentation (check Restricted Permit, BPA classification card, etc)
    (b) Outline of training syllabus
    (c) Routine safety instructions to be observed with aircraft (crossing runways, etc)
    (d) Orientation flight (if desired).

2. Introduction (30 minutes)
    (a) Safety regulations
    (b) Equipment and dress
    (c) Introduction to aircraft to be used in training
    (d) Wind drift determination
    (e) Exit technique (stability)
    (f) Emergency procedures
    (g) Canopy handling
    (h) Landing techniques
    (j) Parachute packing.

3. Familiarisation with Parachutes (90 minutes)
    (a) The anatomy of the main assembly
    (b) The anatomy of the reserve assembly
    (c) The functioning of main and reserve parachutes
    (d) Parachute fitting
    (e) Pre-planning a parachute descent
    (f) Equipment checking procedure.

4. Familiarisation with Aircraft (30 minutes)
    (a) Safety checks
    (b) Procedures for entering and exiting with particular reference to guarding reserve parachutes
    (c) Static line procedure
    (d) Signals and words of command in the air.

5. Aircraft Exits (60 minutes)
    (a) Preparatory commands and signal and actions
    (b) Move into exit position
    (c) Position after exit (stable position)
    (d) Counting, count follow-through and later, dummy ripcord pulls (DRCP).

6. *Emergency Procedures* (90 minutes)
    (a) Verbal count-static line
    (b) Verbal count-free fall:
        (1) count prior to ripcord pull
        (2) count after ripcord pull
    (c) Check of main canopy immediately after opening
    (d) Recognition of malfunctions
    (e) Corrective actions:
        (1) total malfunction
        (2) partial malfunctions (stable and spinning)
    (f) Drill period using suspended harness.

7. *Canopy Handling* (60 minutes)
    (using suspended harness if possible)
    (a) Check canopy
    (b) Orientation with ground:
        (1) grasp toggles
        (2) ascertain location over ground, target and drift
        (3) work to wind line (zigzag method, to obtain)
        (4) check vertical angle of descent (hold or run)
        (5) avoidance of obstacles (do not become intent on target)
        (6) suspended harness drill period
    (c) Prepare to land:
        (1) attitude to adopt landing position: approx 150ft (8–10sec)
        (2) body position, face into wind
        (3) obeying ground instructions if loudspeaker equipment is available.

8. *Parachute Landing Falls* (90 minutes)
    (a) Types:
        (1) normal (front, back side)
        (2) tree
        (3) power line
        (4) water
    (b) Five points of body contact
    (c) Recovery from drag:
        (1) hit, roll, recover, run
        (2) pulling lines
        (3) Capewell (harness clips).

9. *Field Rolling the Parachute* (30 minutes)
    (a) Chain lines
    (b) Sleeve over canopy

(c) Close one side flap with pack-opening bands
(d) Secure all equipment and move to packing area.

10. *Dropping Zone Duties* (30 minutes)
    (a) Responsibility
    (b) Control
    (c) Rotation of personnel.

11. *Parachute Packing Instruction* (backpacks only) (180 minutes)

12. *Testing—all phases* (60 minutes)

Whilst the BPA training schedule thus lays down a clear-cut formula for initiating the novice, it is the instructor who brings to life the rules by projecting them in the light of his own experience. His first task is to gain the confidence of his pupil, and his second to impress upon him or her the vital importance of concentration and strict adherence to the basic rules. Each instructor will have his own technique, but the common aim is a safe and competent sky-diver properly trained in accordance with the BPA's well-proven standard of requirements.

MAIN AND RESERVE PARACHUTES

During the initial course all necessary equipment is provided and can be hired on subsequent occasions. Once qualified, however, the sport parachutist will almost certainly wish to acquire his own parachute and while joint ownership arrangements are common in light aircraft, gliding and ballooning clubs, the parachute is an immensely personal item and, fortunately, within the means of the individual.

The price of an ordinary canopy and harness, at the time of writing, is in the region of £200 and, properly cared for, it should last for several years. Steerable, high-performance parachutes are, naturally, more expensive and could cost over £300. On the credit side, there is comparatively little wear and tear on parachuting equipment and depreciation is low.

Parachute canopies are made of nylon in varying degrees of porosity according to type and design. The normal canopy consists of a number of panels, or gores, incorporating shrouds or rigging lines. A 24ft diameter canopy would have 24 gores and shroud lines, each gore comprising four sections of nylon and so confining any tear or damage to a small area. The shroud lines are embedded in the seams joining the gores in the course of construction so that the lines run up from the harness into the canopy and back down to the harness. There they taper down to form the familiar upturned cone-shaped profile and in four groupings meet the connecting links and risers which incorporate the Capewell release, a locking device for clipping the body harness to the risers.

Sport parachutists are required to wear two parachutes on all jumping occasions—usually a main parachute of 28ft diameter worn on the back and a 24ft diameter reserve parachute strapped to the chest. A 28ft canopy bearing a 200lb load will stabilise the rate of fall to roughly 20ft per second and the 24ft 'chute, under a similar load, to about 25ft per second. The main parachute has attached to it a small pilot 'chute which is the first to be ejected from the pack when the ripcord is pulled, so dragging out the main canopy and helping its deployment in the slipstream of the falling body. The reserve, having no pilot 'chute, deploys more rapidly and so saves vital seconds in an emergency.

Accurate folding and packing of the parachute is of vital importance and students are not allowed to undertake this task without supervision until they have reached a high standard of competence and been awarded a packing certificate. It is also a rule of the BPA that a parachute which has remained packed for 90 days or more may not be used until it has been unpacked, aired and repacked.

THE JUMP AIRCRAFT

Few civilian parachutists will have the opportunity of jumping from the large types of aircraft used by Service teams, such as the giant troop-carriers, parachuting from which enables the parachutist to run into his jump and has been compared by a member of the famous Red Devils' squad to 'jumping off the back of a lorry'. Instead, most clubs use a small single-engined craft, which is either their own property or on hire from a nearby flying club.

The major requirement of the aircraft is that it should be capable of maintaining even flight at comparatively low speeds, as jumping usually takes place at between 70 and 80mph. The Cessna range with its high-wing configuration is very popular for this purpose as, with the door removed and a small steel plate placed over a wheel of the fixed undercarriage, the parachutist can move out on to the plate, taking a firm grip on the wing strut and, at a given signal, launch himself into space with little risk of catching himself or his equipment on any part of the aircraft. The Cessna 172, for example, can lift three parachutists, while the slightly larger model can carry four at a time. If fuel costs are to be kept down, the pilot of the jump aircraft must waste no time in reaching the required height and returning to land and, in the hands of a good pilot, an aircraft dropping experienced parachutists from a height of 3,000ft is likely to be back on the ground before the jumpers.

THE SPECIALISTS

For advanced sky-diving, where free-falling enthusiasts perform 'star' or relative work which involves linking manoeuvres, larger aircraft, such as the Britten-Norman Islander and the Short Skyvan, are employed and are capable of lifting nine or more parachutists at a time. The cost of hiring such aircraft can often be offset by income derived from giving displays at agricultural shows and fetes and many British clubs run such specialist demonstration teams. The Red Devils squad are, of course, regular soldiers but, even so, their display activities are mainly self-supporting and when the members are not training, performing army duties or appearing at displays, they visit schools and clubs to give lectures.

As jumping aircraft have to be approved by the CAA, so pilots must also hold certain qualifications before they are permitted to drop parachutists. They must have flown more than 100 flying hours solo, or in command of an aircraft, and must also have been

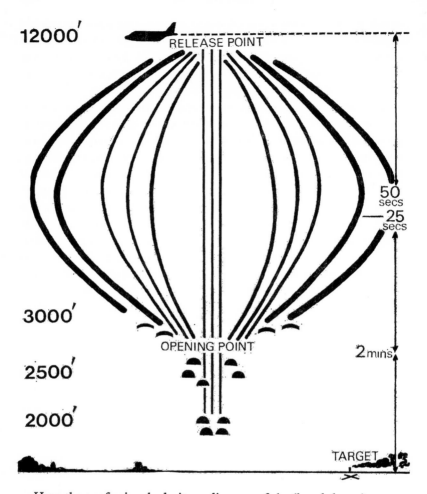

How the professionals do it—a diagram of the 'bomb burst' performed by the Falcons, the RAF's free-fall team. After leaving the aircraft in one group at a height of 10,000–12,000ft, after 5 seconds each man tracks in a different direction for 25 seconds, then turns and tracks back to the opening point. The first six men to jump open their chutes at 2,000ft and the remainder at 2,500ft. The team then guide their parachutes to land on a target area approximately 20yd in diameter

tested by a parachute instructor, who countersigns the pilot's application for certification. The pilot is always in command of the aircraft but when students or intermediate jumpers are being carried there will be a jumpmaster aboard and only on his signal may anyone leave the aircraft.

Qualifications for an instructor approved by the BPA are no less rigid and, in addition to holding a Category X certificate in proof of his extensive parachuting experience, a candidate must have made more than 100 delayed opening descents, of which at least 25 were from a height of at least 7,000ft. No less important will be the qualifications inherent in his own personality: judgment, ability to instil confidence and to detect any flaws in the temperament of a novice, a keen sense of responsibility, and the ability to talk to the student in layman's language. One illustration of the instructor's own advanced training is a test of his ability to deal with a 'hang up'. This is a situation where the student's static line has failed to open the parachute and he is being towed along by the aircraft. The jumpmaster, always equipped with a knife, signals his intention to saw through the line to the student, who waits, hand on reserve parachute release, to be cast adrift before deploying his reserve canopy.

Specialisation in parachuting takes three different forms. First comes style during a free-fall, the period between leaving the aircraft and the manual opening of the canopy. Secondly comes accuracy in landing and thirdly the increasingly popular 'star' or 'relative work'. For the style sections there are conventional manouvres laid down, such as holding a stable position, making turns to left or right, and barrel rolls.

Accuracy in landing involves spotting, or planning a descent before leaving the aircraft. Once the parachute is open the jumper follows the line of his planned course towards the ground target being 'attacked'. This is usually an area of fine gravel some 30 metres in diameter, with a crossline target and a disc in the centre. The parachutist scores points according to the point of impact, measured from the centre. When dead centre is the aim, landing style comes second and the competitor concerns himself only with getting a heel as close as possible to the bullseye.

G

In 'star' or 'relative work', several parachutists combine in free fall to form patterns and link-ups, before dispersing to deploy their canopies. (Picture, p. 72.)

CHAMPIONSHIPS AND COMPETITIONS

The world parachute championship, held biennially, consists of free-fall competitions in both style and accuracy categories and in 1974 Britain was, for the first time, in a position to enter a women's team of the required calibre and experience. An innovation in 1975 will be the first of a biennial series of world 'relative work' championships, in which there will be two basic competitions. One will be for a team of ten, which will be required to link hands and form a star, the winning team being the one to achieve this formation in the shortest time. The second will be for teams of four, each of which will be required to carry out eight specific team manoeuvres in the course of one free-fall. 'Relative' work, consisting as it does of teams of parachutists manoeuvring relative to each other while free-falling, is certainly the most spectacular form of free-fall parachuting and well deserving of its own world championship.

At a less rarified level, local club competitions are frequently held in Britain under BPA rules and parachutists travel long distances to take part. On such occasions parachuting may start at breakfast time and go on till sundown, with perhaps a barbecue in the evening staged by the club members if it is a weekend competition. The venue may be in a rural area where a club may have a dropping zone, a landing strip and one or more light aircraft of its own. To maintain a tight jumping schedule, two light aircraft may be in use, and if free-falling is being practised, multi link-ups can be achieved by using both aircraft flying on parallel courses at different heights. The jumpers follow a joint plan, based on calculations of each man's weight, and by a pre-determined sequence of delayed openings steer themselves together to join up into six- or eight-man star formations.

Should a jumper find himself out of control in a free fall and unable to regain it by body manoeuvres, BPA regulations demand that the main parachute must be deployed at once. In any event, canopies must be opened within 2,000ft of the ground, with the exception of display teams who are allowed to deploy down to 1,500ft. Parachuting is prohibited through cloud—so that a cloud base of 2,300ft or so generally precludes jumping—and in free falls in company the lower man always has the right of way. All static-line descents must be made from a height of not less than 2,500ft, though 'jumps and pulls' may be carried out down to a minimum of 2,200ft.

The highest of the various grades of competence devised by the BPA is Category X. This is the grade required of an approved instructor and among its privileges is that of being permitted to use a camera in free fall—with the opportunity of securing spectacular photographs of free-falling parachutists against the chequered background of the landscape, such as that reproduced on page 71. Not everyone aspires to Category X qualifications but the keen sport parachutist should have no difficulty in eventually qualifying for a General Permit at Category VIII, which gives him sole responsibility for the condition of his equipment and entitles him to operate free from direct supervision by a BPA instructor.

Whatever it be that the sport parachutist decides to specialise in—whether it be free fall style, accuracy jumping or a specialised form of aerial photography—he or she can either operate as an individual or join with others in highly developed teamwork. But the era of the pioneer who could risk his neck without let or hindrance ended quietly with the publication of this order nearly 50 years ago:

> It is notified under Article 13 of the Air Navigation (Consolidation) Order 1920, as amended by the Air Navigation (Amendment), 1925 that parachute descents from civil aircraft are prohibited unless permitted by directions issued by the Secretary of State. Formal Application for the necessary permission giving full details of the proposed descent, should be addressed to The Secretary (A & I) Air Ministry, Gwydyr House, Whitehall, S.W.1, at least fourteen days prior to the date on which it is desired to make the parachute descent.

The day of controlled parachuting had arrived.

PARASCENDING

Parascending, the launching from the ground of a parachutist towed behind a vehicle in the same way as a conventional glider, owes much of its rapid growth in the last few years to several factors. There is no aircraft requirement; operations can be carried out well below the cloud levels that would normally preclude parachuting; training and practice can be supervised and finely controlled by the instructor in the towing vehicle. And, subject to certain safety regulations, flights over water can be achieved by power-boat tow.

Parascending largely came about through the work of a Frenchman, P. M. Lemoigne, who designed and developed a parachute that would fly forward using air flow and also be controllable using a soft-wing glider principle. In 1964, the American-based Pioneer Parachute Co, using the Lemoigne principle, developed the Para-Commander as a world sport championship canopy. It measures 24ft in diameter and can achieve a forward speed of about 12mph.

As a flying machine, the Para-Commander has both lift and drag. It is also highly manoeuvrable and in four seconds it is possible to execute a 360 degree turn. A number of slots are built into the canopy, which has skirts both to left and right. Air is thereby allowed to pass over the canopy's contours in a similar way to an aerofoil section or aircraft wing. There is a lift and drag ratio of 1·16, meaning 11·6ft horizontal travel for every 10ft vertical. By a system of toggles the parachute can be steered and controlled with considerable accuracy.

This comparatively new sport developed under the umbrella of the British Parachute Association until 1973, when the offspring became independent as the Association of Parascending Clubs. The new association inherited from its parent an established framework of training procedure and rules. There are now about 80 member clubs in the APC. Although there are many Service, Territorial and Army Cadet force clubs, some 28 civilian and 5 scouting units are also active.

Clubs rather than individual members comprise the association. There is a £5 annual membership and £1 training fee. Flight costs are applied in various ways. Some clubs divide equally the session expenses, others charge so much per flight. As an indication, scouting clubs quote 40p a flight, although a weekend course at advanced level may cost members nearer £7 all-in.

There are no health certificate requirements and, indeed, no junior age limit to flying. Parents' permission is necessary for those under 18 years and the scouting movement will not permit under 14s unattached descent. A member is not required to provide a parachute—it costs about £200—as the club usually has two sets of equipment, overalls and helmets on hand. All the member needs to provide is a pair of good ankle boots, although parachute boots with pneumatic soles are recommended.

Out on the field the parachutist faces into wind and, attached by his harness through a quick-release device and yoke, is linked to the towline. The towing vehicle is usually a Land-Rover or other field car, travelling upwind. The canopy is carefully laid out downwind of the parachutist.

When all is ready for the launch, a marshal signals the driver by pre-arranged code, two canopy wing-tip-holding assistants lift sections of the parachute and the wind fills the canopy. When the towline takes up the weight, the parachutist moves forward and is soon off the ground. During ascent the car driver controls the rate of climb by his own speed. (Picture, p 36.)

The proficient parascender will normally be towed until reaching about 800ft before releasing himself. There is no extensive training programme involved and the newcomer can reach this standard within a weekend or two. The basic training schedule follows these exercises:

1. Landing roll.
2. Inflation and collapse of canopy by canopy holders.
3. Collapse of canopy by parachutist on ground.
4. Briefing before first towed flight.
5. Very slow descents controlled by driver with release. One or two long flights.
6. Slow, medium and full rate descents controlled by driver without release. Six full rate.

7. Canopy control practice by rear lift webs and by steering toggles on the ground. Repeat.
8. First ascent to 200–300ft. Self release. No turns. One or two ascents.
9. Ascents to 500–800ft. Self release. 90 degree turns. 3-plus ascents.
10. Ascents to 500–800ft. Self release. Gentle turns. About ten flights.
11. Ascents to 500–800ft. Self release. Canopy instability. Five to ten flights.
12. Ascents to 500–800ft. Self release. Turns and target. Five to ten flights.

A number of further flights are required to master the techniques of braking, stall and recovery, and generally perfecting flying ability under varying wind conditions. Wind gusts rather than wind force present problems, 12mph being most acceptable in assisting ascents. The final part of the training schedule is 'target' approach and flying back from 'release' to land on a 10cm disc.

Parascending as a sport is now well established and twenty-three clubs entered teams for the 1974 National Championships held at Watchfield airfield, Oxford.

WHERE TO FLY

The following guide to flying-training schools and clubs in the United Kingdom and Eire is the joint compilation of *Pilot* magazine and the British Light Aviation Centre and is here reproduced by their kind permission. Clubs are listed in this order: England, by counties in alphabetical order; followed by the Channel Islands, Scotland, Wales, Isle of Man, Northern Ireland and Eire.

ABBREVIATIONS

SPL, PPL, CPL: Student-, Private-, Commercial-Pilot's Licences. IMC: Instrument Meteorological Conditions (rating). I/R: (full) Instrument Rating. FAA: the US Federal Aviation Administration. R/T: Radio Telephony (rating). D4 LINK, FRASCA and ANT 18 are Flight Simulators.

* Indicates clubs that operate CAA approved courses

BEDFORDSHIRE
*Luton Flying Club
Luton Airport, Bedfordshire
Courses: PPL, Night, IMC, Twin,
Instrument, Instructor, CPL/IR
***Rogers Aviation International
Flight School**
c/o Rogers Aviation Ltd,
Cranfield Aerodrome,
Great Barford, Bedfordshire
Courses: PPL, Night, IMC, Twin,
Instructor, Aerobatics, PPL/IR, R/T
(FAA PPL, CPL, IR)

BERKSHIRE
*West London Aero Club
White Waltham Aerodrome,
Maidenhead, Berkshire
Courses: PPL, Night, IMC, Twin,
Instructor, Aerobatics, FRASCA

BUCKINGHAMSHIRE
*Gregory Flying Training School
Denham Aerodrome,
Buckinghamshire
Courses: PPL, Night, IMC, TWIN,
Aerobatics, Instructor, Helicopter,
R/T
Lapwing Flying Group
c/o Mrs P. Ward,
Docemonte, Abbess Roding,
Ongar, Essex
Courses: PPL, Night, IMC, Twin,
Aerobatics
*Airways Flying Club
Wycombe Air Park,
Booker, Marlow, Buckinghamshire
Courses: PPL, Night, IMC, Instructor,
R/T
*Wycombe Air Centre Ltd
Wycombe Air Park,
Booker, near Marlow,

Buckinghamshire
Courses: PPL, Night, IMC, Twin,
Instructor, I/R renewals
Simulated Flight Training Ltd
Wycombe Air Park,
Booker, Marlow, Buckinghamshire
Kelly Aeroplane Ltd
Wycombe Air Park,
Booker, Marlow, Buckinghamshire
Course: Aerobatics for PPL holders

CAMBRIDGESHIRE
*Cambridge Aero Club
Cambridge Aerodrome,
Newmarket Road, Cambridge
Courses: PPL, Night, IMC, Twin, D4
LINK

CHESHIRE
Cheshire Air Training School
No 3 Hangar, Liverpool Airport,
Cheshire
Courses: PPL, Night, IMC, Aerobatics,
R/T, Instructor, I/R

CORNWALL
Bodmin Flying Club
Near Bodmin, Cornwall

CUMBERLAND
*Oxford Air Training School
Carlisle Airport,
Cumberland CA6 4NW
Courses: PPL, CPL, Night, IMC, Twin,
Instructor, I/R, Aerobatics, R/T

DERBYSHIRE
Merlin Flying Club Ltd
PO Box 31,
Derby
Courses: PPL, Night, IMC
*East Midland School of Flying
East Midland Airport,
Castle Donington, Derbyshire
Courses: PPL, Night, IMC, Instructor,
Aerobatics
Alidair Flying Club
East Midland Airport,
Castle Donington, Derbyshire
Courses: PPL, IMC, Night, Aerobatics

DEVON
*Plymouth Aero Club
Plymouth Airport,
Crownhill, Devon

Courses: PPL, Night, IMC, Instructor
Dunkeswell Aero Club
Dunkeswell Aerodrome,
Near Honiton, Devon
Courses: PPL, Night, IMC
Exeter Flying Club
The Airport, Exeter, Devon
Courses: PPL, Night, IMC, R/T

DURHAM
Sunderland Flying Club Ltd
Sunderland Airport,
Washington Road,
Sunderland, Co Durham
Courses: PPL, Night, IMC, Aerobatics
Middleton St George Aero Club
Teesside Airport,
Darlington, Co Durham
Courses: PPL, Night, IMC, Aerobatics

ESSEX
Stapleford Flying Club Ltd
Stapleford Aerodrome,
Near Romford, Essex
Tel: Stapleford 380
Courses: PPL, Night, IMC, Twin, R/T
*Southend Light Aviation Centre
Southend Airport, Essex
Courses: PPL, Night, IMC, Twin,
Instructor, Aerobatics
Southend Aero Club
Southend Airport, Essex
Courses: PPL, Night, IMC, Twin,
Aerobatics, R/T
Skywork Ltd
Building 52,
Stansted Airport, Essex
Courses: PPL, IMC, Night, Twin,
Instrument, Aerobatics
Dengie Hundred Flying Club
Witchards, Scotts Hill,
Southminster, Essex
Courses: PPL, Night, IMC, Twin
Herts & Essex Air Services
Stapleford Aerodrome,
Near Romford, Essex RM2 1RL
Courses: PPL, Night, IMC, Twin, I/R

GLOUCESTERSHIRE
APC Flying Group
King Road Avenue,
Avonmouth, Bristol
Cotswold Aero Club

Staverton Airport,
Cheltenham, Gloucestershire
Courses: PPL, Night, IMC, Instrument,
Aerobatics, R/T
Bristol & Wessex Aeroplane Club
Bristol (Lulsgate) Airport,
Bristol BS19 3DS
Courses: PPL, Night, IMC, R/T
Staverton Flying School
Staverton Airport,
Cheltenham, Gloucestershire
Courses: PPL, Night, IMC, Aerobatics,
ANT 18

HAMPSHIRE
*****Western Air Training Ltd**
Thruxton Aerodrome,
Andover, Hampshire
Courses: PPL, Night, IMC, Instructor,
Twin Aerobatics
Portsmouth Flying School
City Airport,
Portsmouth, Hampshire PO3 5PD
Courses: PPL, IMC, Night, Aerobatics,
R/T
*****Southampton School of Flying**
Southampton Airport, Hampshire
Courses: PPL, Night, IMC, Twin,
Instrument, Aerobatics
Danebury Flying Group
Danebury, Barley, Hampshire
Courses: PPL, Night, IMC, Aerobatics
Cannon Aviation School of Flying
Hurn Airport,
Bournemouth, Hampshire
Courses: PPL, Night, IMC, Assistant
Instructor's
Condor Flying Club
Thruxton Aerodrome,
Andover, Hampshire
Courses: PPL, Night, Aerobatics
Vectair Flying Club
Portsmouth City Airport, Hampshire
*****Bournemouth Air Centre**
Hurn Airport, Hampshire
College of Air Training
Hamble, Hampshire
(Trains pilots for British Airways)

HEREFORDSHIRE
Herefordshire Aero Club
c/o Wing Cdr A. James, Sec,
Kingsland, Near Leominster,

Herefordshire HR6 9QN
Courses: PPL, Night, IMC, R/T

HERTFORDSHIRE
Val Air Flying Group
39 Western Road,
Tring, Hertfordshire
Courses: Night, IMC
*****London School of Flying**
Elstree Aerodrome,
Boreham Wood, Hertfordshire
Courses: PPL, Night, IMC, Twin,
R/T, LINK D4

KENT
Air Touring Club
Biggin Hill Airport,
Westerham, Kent
Courses: PPL, Night, IMC, Aerobatics,
R/T
Alouette Flying Club
Biggin Hill Aerodrome,
Westerham, Kent
Courses: PPL, Night, IMC, Aerobatics
Civil Service Flying Club
Biggin Hill Airfield, Kent
Courses: PPL, IMC, Night
Anderson Flying School
Biggin Hill Airfield, Kent
Courses: PPL, Night, IMC, R/T
West Essex Flying Club
Biggin Hill Airfield, Kent
Courses: PPL, Night, IMC, Twin
Flairavia Flying Club
Biggin Hill Airfield,
Westerham, Kent
Courses: PPL, Night, IMC, Aerobatics,
R/T
Experimental Flying Group
56 Haydn Avenue, Purley,
Surrey CR2 4AF
Courses: PPL, Night, IMC, Aerobatics
*****Surrey & Kent Flying Club**
Biggin Hill Airport,
Westerham, Kent
Courses: PPL, Night, IMC, Twin,
Instrument, Instructor, Aerobatics,
LINK D4
Experimental Flying Group
Biggin Hill Airport,
Westerham, Kent
(weekends and evenings only)
Courses: PPL, Night, IMC, Aerobatics

Sportair Flying Club
Biggin Hill Aerodrome,
Westerham, Kent
Courses: PPL, Night, IMC, Aerobatics,
R/T
South London Aero Club
Biggin Hill Airport, Kent
Courses: PPL, Night, IMC, Twin
Biggin Hill Flying Club
Biggin Hill Airport,
Westerham, Kent
Courses: PPL, Night, IMC, Aerobatics
King Air Flying Club
Buildings 160-2,
Biggin Hill Airport,
Westerham, Kent
Courses: PPL, Night, IMC, Twin,
Instrument, Aerobatics, R/T
Cinque Ports Flying Club
c/o Business Air Travel Ltd,
Ashford (Lympne) Airport,
Hythe, Kent
Courses: PPL, Night, IMC
Rochester Flying Club
Rochester City Airport, Kent
Courses: PPL, Night, IMC, I/R,
Aerobatics, LINK D4
Rye & Romney Flying Club
Lydd Airport,
Romney Marsh, Kent
Courses: PPL, Night, IMC, I/R,
Aerobatics, R/T
Headcorn Flying School
Headcorn Aerodrome, Kent
Courses: PPL, Night, IMC, Aerobatics

LANCASHIRE
West Lancashire Aero Club
(RAF Woodvale)
c/o Miss J. Griffith,
Bragg & Collins Ltd,
11-13 Fazakerly Road, Walton,
Liverpool L9 1BX
Course: PPL
Woodvale Aero Club
RAF Woodvale,
1B Belgrave Road,
Southport, Lancashire
Courses: PPL, Night, IMC
**Air Navigation, Trading Co Ltd,
Flying Club**
Blackpool Airport,

Lancashire FY4 2QS
Courses: PPL, Night, IMC, Aerobatics,
R/T
Southport & Merseyside Aero Club
Building No 6,
Liverpool Airport, Lancashire
Courses: I/R, Night, IMC
**Liverpool Aero Club & Flying
School**
Hangar 4,
Liverpool Airport,
Speke, Liverpool L24 877
Courses: PPL, Night, IMC, Twin
Blackpool, Fylde Aero Club
Blackpool Airport, Lancashire
Courses: PPL, Night, IMC
*Lancashire Aero Club
Barton Aerodrome,
Eccles, Lancashire M30 7SA
Courses: PPL, IMC, Aerobatics, R/T
***Northern School of Aviation Flying
Club**
Blackpool Airport, Lancashire
Courses: PPL, Night, IMC, Twin,
Instructor, Aerobatics, R/T
Manchester Advanced Flying Group
Hangar 522,
Ringway Airport, Manchester
Courses: PPL, Night, IMC, Twin, R/T

LEICESTERSHIRE
Leicestershire Aero Club Ltd
Leicester East Airfield,
Gartree Road, Leicester LE2 2FG
Courses: PPL, Night, IMC, Twin,
Aerobatics, Instructor, R/T

LINCOLNSHIRE
Wickenby Flying Club
Wickenby Airfield, Near Wragby,
Lincoln, Lincolnshire
Courses: PPL, Night, IMC, RT,
Aerobatics
Fenland Aero Club
Holbeach St Johns,
Spalding, Lincolnshire
Course: PPL
*Skegness Aero Club
Ingoldmells, Skegness, Lincolnshire
Courses: PPL, IMC, R/T

LONDON
Graeme Percival
Room 129, Grosvenor Hotel,
London SW1
Course: Gat 1 Simulator

NORFOLK
Norfolk, Norwich Aero Club Ltd
The Clubhouse,
Swanton Morley,
Dereham, Norfolk
Courses: PPL, Night, IMC, R/T,
Aerobatics

NORTHAMPTONSHIRE
*Peterborough Aero Club
Sibson Aerodrome,
Peterborough,
Northamptonshire PE8 6NE
Courses: PPL, Night, IMC, Instructor,
Aerobatics
*Northamptonshire Aero Club
Sywell Aerodrome,
Northampton, Northamptonshire
Courses: PPL, Night, IMC, Aerobatics,
R/T

NOTTINGHAMSHIRE
Sherwood Flying Club
18 Ridge Lane,
Radcliffe on Trent, Nottinghamshire
Courses: PPL, Night, IMC, Aerobatics,
R/T
Sheffield Aero Club
Netherthorpe Airfield,
Thorpe Salwin,
Near Worksop, Nottinghamshire
Courses: PPL, IMC, R/T, ANT 18

NORTHUMBERLAND
Border Flying Group
c/o Colin Clark Esq,
'Three Ways',
Norham, Berwick-on-Tweed,
Northumberland
Courses: PPL, Night, IMC
Newcastle upon Tyne Aero Club
North East Airport,
Woolsington, Newcastle upon Tyne
Courses: PPL, Night, IMC, R/T, D4
LINK

OXFORDSHIRE
*Oxfordshire Air Training School
Oxford Airport,
Kidlington, Oxford
Courses: PPL, CPL, Night, IMC, Twin,
Instructor, I/R, Aerobatics

SHROPSHIRE
Shropshire Aero Club
c/o Mr V. G. Beaumont,
9 St Marys Street,
Shrewsbury, Shropshire
Courses: PPL, Night, IMC, Aerobatics

SOMERSET
Achilles School of Flying
Weston Airport,
Weston Super Mare, Somerset
Courses: PPL, Night, IMC, Twin,
Instructor, Aerobatics, R/T, D4 LINK

SUFFOLK
British School of Flying
Ipswich Airport, Suffolk
Courses: PPL, Night, IMC, Twin,
Instructor, Aerobatics, CPL/IR, R/T
Ipswich School of Flying
Ipswich Airport, Suffolk
Courses: PPL, Night, IMC

SURREY
Aeromart Flying Club
Blackbushe Airport,
Camberley, Surrey
Courses: PPL, Night, IMC, Twin
Aerobatics International Ltd
62 Ennerdale Road,
Kew, Richmond, Surrey TW6 2DL
Course: Aerobatics
Eurotec Flying Group
68 Stratton Road,
Sunbury-on-Thames, Middlesex
Aerodrome: Blackbushe
Courses: PPL, Night, IMC
Tiger Club
Redhill Aerodrome,
Redhill, Surrey
Courses: Night, IMC, Aerobatics
(Must have 100 hours to join)
Three Counties Aero Club
Blackbushe Airport,
Camberley, Surrey

Courses: PPL, Night, IMC, Twin,
Aerobatics, I/R, R/T, FRASCA
*Fair Oaks School of Flying
Fair Oaks Aerodrome,
Chobham, Surrey
Courses: PPL, Night, IMC, Twin,
Instructor, Aerobatics, R/T
Fair Oaks Aero Club
Chobham, Near Woking, Surrey
Courses: PPL, Night, IMC, Twin,
Instructors, Aerobatics, R/T, D4
LINK
Blackbushe Aero Club
Blackbushe Airport,
Camberley, Surrey
Courses: PPL, Night, IMC, R/T
Mack Airways Training Ltd
Gatwick Airport, Surrey
Course: I/R, Simulators
***London Transport Sports Flying
Club**
Fairoaks Aerodrome,
Chobham, Surrey
Monarch Aero Club
Blackbushe Airport,
Camberley, Surrey
Courses: PPL, Night, IMC, R/T

SUSSEX
Shoreham School of Flying
Shoreham Airport, Sussex
Courses: PPL, Night, IMC,
Aerobatics, R/T
Goodwood Flying School
Goodwood Aerodrome,
Chichester, Sussex
Courses: PPL, Night, IMC, Twin,
Instructor, I/R, Aerobatics
Brighton Flying Group
Shoreham Airport, Sussex
Courses: PPL, Night, IMC, Aerobatics,
R/T

WARWICKSHIRE
Coventry Aeroplane Club
Baginton, Coventry, Warwickshire
Courses: PPL, Night, IMC, R/T
**Sir W. G. Armstrong Whitworth
Flying Group**
Coventry Aerodrome, Warwickshire
Courses: PPL, Night, IMC, R/T
Warwickshire Aero Club
Birmingham Airport,

Birmingham 26, Warwickshire
Courses: PPL, Night, IMC, Instrument,
Instructor
Executive Air (Medminster) Ltd
Birmingham Airport,
Birmingham 26, Warwickshire
Courses: PPL, Night, IMC, Aerobatics
R/T
Skeldor Flying Group
Coventry Airport, Warwickshire
Courses: PPL, Night, IMC
Coventry Flying Club
Coventry Airport,
Baginton, Warwickshire

WILTSHIRE
**Dorset Flying Club & Aviation
Centre**
Compton Abbas Airfield,
Ashmore, Salisbury, Wiltshire
Courses: PPL, Night, IMC, Twin, I/R,
R/T
Bustard Flying Club Ltd
c/o A & AEE, Boscombe Down,
Amesbury, Wiltshire
Courses: PPL, Aerobatics

WORCESTERSHIRE
*Midland Aviation Centre
Halfpenny Green Aerodrome,
Bobbington, Near Stourbridge,
Worcestershire
Courses: PPL, Night, IMC, R/T, Twin,
Instructor

YORKSHIRE
*Yorkshire Aeroplane Club
Leeds Bradford Airport,
Leeds LS19 7TU
Courses: PPL, Night, IMC, Twin,
Instructor, Aerobatics, R/T, LINK
Hull Aero Club Ltd
c/o W. K. Charles Esq,
14 High Street,
North Ferriby, East Yorkshire
HU14 3JP
Courses: PPL, Night, IMC, Twin,
Areobatics
*Sherburn Aero Club Ltd
Sherburn-in-Elmet Airfield,
Leeds, Yorkshire
Courses: PPL, Night, IMC, Instructor,
Aerobatics

*Doncaster Aero Club
Control Tower,
Doncaster Airport,
Doncaster, Yorkshire
Courses: PPL, Night, IMC, Twin,
Instructor, Aerobatics
Northair
Leeds/Bradford Airport, Yorkshire

CHANNEL ISLANDS
*Channel Islands Aero Club
The Airport, St Peter, Jersey, CI
Courses: PPL, Night, IMC, Instructor,
Aerobatics
**Channel School of Flying
(Guernsey) Ltd**
States Airport, Guernsey, CI
Courses: PPL, Night, IMC
Galleon Flying Group
Les Heches,
St Peters in the Wood,
Guernsey, CI
Courses: PPL, Night, IMC, Twin, R/T

SCOTLAND
Pegasus Flying Club
Dyce Airport, Aberdeen, Scotland
Courses: PPL, Night, IMC, R/T
**Strathspey Aviation Co Ltd
Flying Club**
15–19 High Street,
Aberlour-on-Spey, Banffshire,
AB3 9QL, Scotland
Courses: PPL, Night, IMC, I/R, R/T
Glasgow Flying Club
Glasgow Airport, Scotland
Courses: PPL, Night, IMC, Aerobatics,
R/T, ANT 18
*Edinburgh Air Centre
Edinburgh (Turnhouse) Airport,
Scotland
Courses: PPL, Night, IMC, Aerobatics,
Twin, R/T
*Air Service Training
Private Flying Centre,
Perth Aerodrome, Perth, Scotland
Courses: PPL, CPL, Night, IMC, Twin,
Instructor, Instrument, Aerobatic,
R/T, D4 LINK
Edinburgh Flying Club
17 St Clair Terrace,
Edinburgh EH10 5NN
Courses: PPL, Night, IMC, Aerobatics

West of Scotland Flying Club
Glasgow Airport, Scotland
Courses: PPL, Night, IMC, Aerobatics
Tayside Flying Club
36 Erskine Terrace,
Monifieth by Dundee, Scotland
Courses: PPL, Night, IMC
Aberdeen Aero Club
Dyce Airport, Scotland
Courses: PPL, Night, IMC, R/T
Orkney Flying Club
Kirkwall, Airport, Orkney
Course: PPL

WALES
South Wales Flying Club
Glamorgan (Rhoose) Airport,
Glamorgan, Wales
Courses: PPL, Night, IMC, Aerobatics
Swansea Flying Club
Swansea Airport,
Glamorgan, Wales, SA5 4QA
Courses: PPL, Night, IMC, Aerobatics,
R/T
West Wales Flying Club Ltd
Swansea Airport, Swansea,
Glamorgan, Wales, SA5 4QA
Courses: PPL, Night, IMC, Aerobatics,
R/T
*Glamorgan Flying Club
c/o Glamorgan (Rhoose) Airport,
Barry, Glamorgan, Wales
Courses: PPL, Night, IMC, Aerobatics,
Instructor, R/T, D4 LINK
Pegasus School of Flying
Rhoose Airport,
Glamorgan, Wales
Courses: PPL, Night, IMC, I/R, Twin

ISLE OF MAN
Manx Flyers Aero Club
Ronaldsway Airport,
Douglas, Isle of Man
Courses: PPL, Night, IMC, Instrument,
Aerobatics, Twin, Instructor, R/T
Isle of Man Flying Centre
Ronaldsway Airport, Isle of Man
Courses: PPL, Night, IMC, Twin,
Instrument/Renewal Aerobatics,
Instructor, R/T

NORTHERN IRELAND
Ulster Flying Club (1961) Ltd
Newtonwnards Airport,
Co Down, Northern Ireland
Courses: PPL, IMC, Aerobatics, D4
LINK
Eniskillen Flying Club Ltd
St Angelo Airport,
Eniskillen, Co Fermanagh,
Northern Ireland
Courses: PPL, IMC, R/T
*****Woodgate Aviation**
Aldergrove Airport, Belfast,
Co Antrim, Northern Ireland
Courses: PPL, Night, IMC, Twin,
Instructor, Aerobatics, R/T

EIRE
Munster Aero Club Ltd
Cork Airport, Irish Republic
Courses: PPL, Night
Dundalk Aero Club
c/o Fairways Hotel, Dundalk,
Co Louth, Irish Republic
Courses: PPL, R/T
Abbeyshrule Aero Club
Co Longford, Irish Republic
Courses: PPL, R/T
Kerry Aero Club
Farranfore, Co Kerry,
Irish Republic
Courses: PPL, IMC, Instructor, R/T

WHERE TO GLIDE

* Indicates clubs that operate at weekends only

Airways Flying Club (Gliding Section) (British Airways staff only)
Wycombe Air Park,
Booker Airfield,
Near Marlow, Buckinghamshire

*Albatross Gliding Club
Davidstow Airfield,
Camelford, Cornwall

*Angus Gliding Club
Royal Naval Air Station,
Arbroath, Angus, Scotland

Aquila Gliding Club
Rutland House,
Ratley, Oxfordshire

*Avro Gliding Club (restricted membership)
c/o Hawker Siddeley Aviation Ltd,
Woodford,
Stockport, Bramhall, Cheshire

*Bath & Wilts Gliding Club
Keevil Aerodrome,
Near Melksham, Wiltshire

*Blackpool & Fylde Gliding Club
Cock Hill Farm,
Fiddlers Lane,
Chipping,
Near Preston, Lancashire

*The Borders (Milfield) Gliding Club
Milfield Aerodrome,
Near Wooler,
Northumberland

Bristol & Gloucestershire Gliding Club
Nympsfield,
Near Stonehouse, Gloucestershire

*Buckminster Gliding Club
Saltby Airfield,
Saltby, Leicestershire

*Burton & Derby Gliding Club
Church Broughton,
Near Foston, Derbyshire

Cairngorm Gliding Club
Feshie Airstrip,
Blackmill Farm
Kincraig,
Inverness-shire, Scotland

*Cambridge University Gliding Club
Cambridge Airport,
Newmarket Road,
Cambridge

Cornish Gliding Club
Trevellas Airfield,
Near Perranporth, Cornwall

*Cotswold Gliding Club
Aston Down Airfield,
Stroud, Gloucestershire

Coventry Gliding Club
Husbands Bosworth Airfield,
Near Rugby, Warwickshire

*Cranfield Institute of Technology GC
Cranfield, Bedfordshire

*Deeside Gliding Club
Aboyne Airfield,
Dinnet,
Aberdeenshire, Scotland

Defford Aero Club
RRE Air Station,
Pershore, Worcestershire

Derby & Lancs Gliding Club
Camphill,
Great Hucklow,
Tideswell, Derbyshire

*Devon & Somerset Gliding Club
North Hill Airfield,
Broadhembury,
Honiton, Devon

Devonshire Soaring Club
Dunkeswell Aerodrome,
Honiton, Devon

Doncaster & District Gliding Club
The Airport,
Ellers Road,
Bessacarr,
Doncaster, Yorkshire

*Dorset Gliding Club
Tarrant Rushton Airfield,
Near Blandford Forum, Dorset

*Dumfries & District Gliding Club
Glaisters,
Kirkgunzeon,
Kirkcudbrightshire, Scotland

*Enstone Eagles Gliding Club
Enstone Airfield,
Enstone,
Chipping Norton, Oxfordshire

*Essex Gliding Club
North Weald Aerodrome,
Essex

*Essex & Suffolk Gliding Club
Barrads Hall Airstrip,
Whatfield,
Hadleigh, Suffolk

*Glamorgan Gliding Club
2½ miles south-west of Maesteg

*Glasgow & West of Scotland
Gliding Club
Operating at Portmoak,
Kinross,
Kinross-shire, Scotland

*Hambletons Gliding Club
Royal Air Force,
Dishforth,
Boroughbridge, Yorkshire

Herefordshire Gliding Club
Shobdon,
Near Leominster,
Herefordshire

*Highland Gliding Club
Milltown Aerodrome,
by Elgin,
Morayshire, Scotland

Imperial College Gliding Club
Lasham Airfield,
Near Alton, Hampshire

*Inkpen Gliding Club
Inkpen Airfield,
Shalbourne,
Marlborough, Wiltshire

*Islay Gliding Club
Port Ellen Aerodrome,
Isle of Islay

Kent Gliding Club
Squids Gate,
Challock,
Near Ashford, Kent

Kirknewton Gliding Club
c/o Scottish Gliding Union,
Portmoak Airfield,
Kinross,
Kinross-shire, Scotland

*Lakes Gliding Club
Walney Airfield,
Barrow-in-Furness, Lancashire

Lancashire Gliding Club
Strathaven Airfield,
Lanarkshire, Scotland

Lasham Gliding Society
Lasham Airfield,
Near Alton, Hampshire

*Lincolnshire Gliding Club
Bardney Airfield,
Bardney, Lincolnshire

London Gliding Club
Dunstable Downs,
Tring Road,
Bedfordshire

Midland Gliding Club
The Long Mynd,
Church Stretton, Shropshire

*****Newcastle & Teeside Gliding
Club**
Carlton-in-Cleveland,
Middlesbrough, Yorkshire

Norfolk Gliding Club
Tibenham Airfield,
Norwich, Norfolk

*****Northumbria Gliding Club**
Currock Hill,
Near Hedley-on-the-Hill,
Stocksfield,
Northumberland

Norwich Soaring Group
c/o Norfolk and Norwich Aero Club,
Royal Air Force,
Swanton Morley,
East Dereham, Norfolk

*****Ouse Gliding Club**
Royal Air Force,
Rufforth, Yorkshire

Oxford Gliding Club
RAF Weston-on-the-Green,
Near Bicester, Oxfordshire

*****Peterborough & Spalding
Gliding Club**
Crowland Airfield,
Postland,
Peterborough

**Polish Air Force Association
Gliding Club**
Lasham Airfield,
Near Alton, Hampshire

Royal Aircraft Establishment
Farnborough, Hampshire

Scottish Gliding Union
Portmoak Airfield,
Scotlandwell,
by Kinross,
Kinross-shire, Scotland

Scout Association Gliding Club
Lasham Airfield,
Near Alton, Hampshire

*****Southdown Gliding Club**
Pariham,
Near Storrington, W Sussex

H

*****South Wales Gliding Club**
Near Gwernewney, S Wales
(2 miles east of Usk on B4235)

South Yorkshire Gliding Club
Hardwick Hall Farm,
Aston,
Near Sheffield, Yorkshire

*****Staffordshire Gliding Club**
Morridge
Near Leek, Staffordshire

**Stratford-upon-Avon Gliding
Club**
Long Marston Airfield,
Near Stratford-upon-Avon,
Warwickshire

Surrey & Hants Gliding Club
Lasham Airfield,
Near Alton, Hampshire

*****Swindon Gliding Club**
South Marston Airfield,
Swindon, Wiltshire

Thames Valley Gliding Club
Wycombe Air Park,
Booker Airfield,
Near Marlow, Buckinghamshire

Tiger Club Soaring Group
Redhill Aerodrome,
Surrey

*****Trent Valley Gliding Club**
Kirton-in-Lindsey,
Near Gainsborough, Lincolnshire

*****Ulster & Shorts Gliding Club**
Newtownards Airport,
Co Down, Northern Ireland

*****University of Aston in Birmingham
GC**
Long Marston Airfield,
Long Marston, Warwickshire

*****Universities of Glasgow &
Strathclyde GC**
Couplaw Farm,
Stratheven,
Lanarkshire, Scotland

*****Upward Bound Trust Gliding
Club**
Haddenham Thame Airfield,
Oxfordshire

*Vale of Neath Gliding Club
Min-yr-Awel,
Tramway,
Hirwaun,
Aberdare,
Glamorganshire, S Wales

*Welland Gliding Club
Marshalls Farm,
Careby,
Stamford, Lincolnshire

West Wales Gliding Association
Withybush Airfield,
Haverfordwest,
Pembrokeshire, W Wales

*Wolds Gliding Club
Pocklington Airfield, Yorkshire

Worcestershire Gliding Club
Paske Field,
Bickmarsh,
Bidford-on-Avon, Worcestershire

Wycombe Gliding School
Booker Airfield,
Near Marlow,
Buckinghamshire

Yorkshire Gliding Club
Sutton Bank,
Thirsk, Yorkshire

Army Gliding Association
Army Aviation Centre,
Middle Wallop,
Stockbridge, Hampshire

Royal Naval Gliding and Soaring
Association
RNAS Yeovilton, Somerset

Royal Air Force Gliding and
Soaring Association
Boscombe Down,
Amesbury, Wiltshire

WHERE TO BALLOON

Europa Balloon Training School
344 Wanstead Park Road, Ilford, Essex
or enquire direct to:
British Balloon & Airship Club
Artillery Mansions, 75 Victoria Street, London SW1H 0JD

WHERE TO PARASCEND

Enquiries to:
The Secretary, BM/Parascending, London WC1 6XX
or to:
O. W. Neumark, 2 Churwell Avenue, Heaton Mersey, Stockport, Cheshire

---

WHERE TO HANG-GLIDE

Enquiries to:
The Secretary, The Hang-Gliding Association, 38 Great Smith Street, London SW1

WHERE TO PARACHUTE

Peterborough Parachute Centre
Sibson Airfield,
Peterborough, Huntingdonshire

Eagle Sport Para Centre
Ashford Airport,
Lympne, Kent

The Sport Parachute Centre
Grindale Field,
Bridlington, Yorkshire

RSA Parachute Club
Thruxton Aerodrome,
Andover, Hampshire

WEEKEND CLUBS

British Parachute Club
Headcorn Airfield,
Headcorn, Kent

Brunel University Skydiving Club
c/o Students Union,
Brunel University,
Uxbridge, Middlesex
(For Brunel Students only)

Black Knights Skydiving Centre
Weeton,
Near Blackpool, Lancashire

**Dunkeswell International
Skydiving Centre**
Dunkeswell Airfield,
Near Honiton, Devon

Hereford Parachute Club
Shobdon Aerodrome,
Leominster, Herefordshire

Leeds/Bradford Parachute Club
16 Tyersal Garth,
Bradford 4, Yorkshire

Leeds University Parachute Club
(Leeds Students only)

Lincoln Pathfinders Free Fall Club
63 Weakland Close,
Sheffield S12 4PD, Yorkshire

Manchester Free Fall Club
Tilstock DZ,
Twenlows Hall Farm,
Whitchurch, Shropshire

Manchester Parachute Club
c/o Secretary: A. E. Cooper,
8 Longford Avenue,
Stratford, Manchester

Metropolitan Police Para Club
Nuthampstead Airfield, Hertfordshire

Martlesham Heath Para Club
Martlesham Heath Aerodrome,
Near Ipswich, Suffolk

Midland Parachute Centre
The Airfield,
Bickmarsh, Worcestershire

North West Parachute Centre
Cark Airfield,
Flookburgh,
Near Grange-over-Sands, Lancashire

**Old Warden Flying & Parachute
Group**
c/o Secretary: D. I. Waugh,
27 Dury Road,
Barnet, Hertfordshire

Paraclan Parachute Club
Charter Hall,
Near Berwick

Southern Skydiving Club
Bembridge Airfield,
Isle of Wight

Scottish Parachute Club
Strathallen Castle,
Auchterarder, Perthshire, Scotland

Staffordshire Skydiving Club
c/o Secretary: D. Capper,
61 Hazlehurst Street,
Hanley,
Stoke-on-Trent, Staffordshire

South Staffordshire Skydiving Club
Halfpenny Green Aerodrome,
Bobbington, Worcestershire

Northumbria Parachute Club
Sunderland Airport,
Washington Road,
Sunderland, Co Durham

Queen Mary Skydiving Club
c/o Secretary: Miss A. Kelly,
Lynden Hall,
110 High Road,
Woodford, London E18

Vauxhall Skydiving Club
Halfpenny Green Aerodrome,
Bobbington, Worcestershire
(Vauxhall Motors only)

Warwickshire Aviation & Sport Para School
c/o Secretary: P. A. Howell,
17 Westhill Close,
Olton,
Solihull, Warwickshire

Wessex Parachute Club
Swallow Cliff,
Near Shaftesbury, Dorset

Yorkshire Aeroplane Club Sport Parachute Wing
c/o Secretary: G. C. P. Shea-Simonds,
Yorkshire Flying Services,
Leeds/Bradford Airport, Yorkshire

Yorkshire F/F Club
c/o Secretary: E. Vine,
21 Wellington Road,
Bridlington, Yorkshire

ADDRESSES OF CONTROLLING BODIES AND ORGANISATIONS

Aerodrome Owners' Association Ltd: Concerned with the interests of owners and operators of aerodromes and landing strips. *Address:* Secretary: W. M. Hargreaves, CBE, 19 Mountview Road, London N4 4SS.

Air Training Corps, The: A youth movement, open to boys between the ages of 13 and 17¾, designed to encourage a practical interest in flying. Opportunities are provided to gain air experience, flying scholarships are awarded and gliding courses are available. *Address:* HQ Air Cadets, RAF Brampton, near Huntingdon.

Association of Parascending Clubs: Represents the sport in the UK, imposing strict training methods and operating techniques. Organises national championships. Members include service units and youth organisations. *Address:* Secretary: BM/Parascending, London WC1 6XX.

British Balloon and Airship Club: The representative body for the sport of balloon and airship flying in the UK. *Address:* Artillery Mansions, 75 Victoria Street, London SW1H 0JD.

British Gliding Association: Controls and administers the sport of gliding and soaring in the UK. *Address:* Artillery Mansions, 75 Victoria Street, London SW1H 0JD.

British Helicopter Advisory Board: Representative of all interested in the development and use of rotorcraft. *Address:* Knowles House, Cromwell Road, Redhill, Surrey.

British Insurance Association: Representative body of UK insurance companies. Will supply names of general and specialist insurance companies. *Address:* PO Box 538, Aldermary House, Queen Street, London EC4.

British Light Aviation Centre: An association representative of flying clubs, private aircraft owners, small aerodrome proprietors and competitive flying; also responsible for the Flying Instructors' Examining Panel. *Address:* Artillery Mansions, 75 Victoria Street, London SW1H 0JD.

British Parachute Association: The controlling body of sport parachuting in the UK. Responsible for training methods and granting permits to parachutists. Organises national and international events and selects teams to represent Britain. *Address:* Artillery Mansions, 75 Victoria Street, London SW1H 0JD.

British Women Pilots' Association: Assists women to qualify for pilots' licences and gives advice on training and on opportunities open to women in aviation. *Address:* c/o PO Box 13, London SW1.

Business Aircraft Users Association: Offers advisory and information services for the assistance of business users of aircraft in the UK. *Address:* Artillery Mansions, 75 Victoria Street, London SW1H 0JD.

Civil Service Aviation Association Ltd: Affiliated to the Civil Service Sports Council which grants long-term loans for the purchase by civil servants or their organisations of aircraft and equipment for all forms of sports aviation. *Address:* Secretary: J. T. Walker, Woodlands, Marlow Common, Marlow-on-Thames, Buckinghamshire.

Girl Guide Association (Ranger Guides): Provides opportunities for girls to gain flying experience in light aircraft and gliders and to train for aeronautical 'interest certificates' in such fields as 'air crew', 'glider crew' and 'air hostess'. Minimum entry age is 14. *Address:* 17–19 Buckingham Palace Road, London SW1W 0PY.

Girls Venture Corps (Air Wing): Runs air training schemes for girls aged 13–20 in various parts of the UK and offers a number of scholarships to enable members to qualify for a private pilot's licence. *Address:* 33 St George's Drive, London SW1V 4DH.

Hang-Gliding Association, The: Open to membership world-wide. Advises on types of sailwing, organises events and assists with sites. *Address:* 38 Great Smith Street, London SW1.

Popular Flying Association, The: Encourages amateur construction of ultra-light aircraft and the formation of co-ownership groups and clubs, with the object of reducing the cost of sport flying. *Address:* Terminal Building, Shoreham Airport, Shoreham-by-Sea, Sussex.

Scottish Sport Parachute Association: Scottish counterpart of British Parachute Association. *Address:* 13 Boswell Drive, Kinghorn, Fife.

Scout Association (Air Scouts Section): Offers courses for scouts and scout-leaders ranging from weekend experience flights in light aircraft or gliders, instruction in hot-air ballooning and parachuting, to larger 'air adventure camps' including navigational, map-reading and aircraft maintenance instruction. *Address:* Scout Association HQ, 25 Buckingham Palace Road, London SW1W 0PY.

MODEL CONSTITUTION FOR A FLYING SYNDICATE

Reproduced by courtesy of the Dante Balloon Group

1. *Name:* The Club shall be called the Dante Balloon Group ('the Group').

2. *Object:* The object of the Group shall be the purchase and flying of a Cameron o–65 hot-air balloon and the carrying on of any object reasonably incidental thereto to the intent that all Group members shall in due course be able to obtain a CAA Balloon (Hot-Air) Licence.

3. *Numbers:* The Group shall consist of not more than twelve members, each of the twelve founder-members contributing the sum of one hundred pounds on joining the Group.

4. *Committee:* Subject as appears in these rules the Group shall be managed by a committee of three members comprising a chairman, a secretary and treasurer. The committee shall be elected at the annual general meeting in each year and subject to termination of office by resignation or otherwise shall remain in office until their successors are elected at the annual general meeting next following their election. The retiring members of the committee shall be eligible for re-election.

5. *AGM:* An annual general meeting of the Group shall be held in every year not later than 30 November to transact the following business: (a) to receive and, if approved, adopt a Statement of the Group's accounts to 31 October in that year; (b) to appoint the members of the committee; (c) to deal with any special matter which the committee desire to bring before the members.

Notice convening the annual general meeting shall be sent to the members not less than ten days before the meeting.

6. *General Meetings:* A special general meeting may be convened at any time by the committee, notice of which shall be sent to the members at least ten days beforehand.

7. (A) *Quorum:* Save as mentioned in Rule 14 at committee meetings two shall form a quorum.

Save for the purposes of Rules 16 and 17 at general meetings seven shall form a quorum and a resolution proposed shall be deemed to be passed and binding upon all members of the Group if not less than six members vote in favour thereof. Each member shall have one vote upon every motion and in the case of equality of votes the chairman of the meeting shall have a second or casting vote.

(B) *Chairman of Meeting:* The chairman of the Group or, if he shall not be present, any other committee member shall preside as chairman at every general meeting of the Group. If there is no such chairman then the members present shall choose any one of their number to be chairman of the meeting.

8. *Alteration of Rules:* The members of the Group in general meeting shall have power to alter the rules.

9. *Retirement:* A member wishing to withdraw from the Group must give at least one month's notice in writing to that effect to the secretary. Following the expiry of such notice the member shall not be liable to make any further contribution whether financial or otherwise towards the Group but shall not be entitled to the repayment of any contribution already made.

As soon as is reasonably practical after giving notice the withdrawing member shall offer his share to the Group in quorate general meeting which shall have the option of purchasing the same for a sum equal to the value (calculated as below) of the balloon Registration Mark G–AZIP at the date of the general meeting in question divided by the total number of members forming the Group immediately prior to the meeting together with a sum equal to the value at the date of the meeting of the Group's cash reserves (as defined below) divided by the number of members as aforesaid.

For the purpose of this rule and of Rule 16:

(a) the value of the balloon G–AZIP at a given date shall be that proportion of £1,200 which the unexpired portion at that date of the period from 1 January 1972 to 31 December 1976 bears to the whole of that period. After 31 December 1976 the value of the balloon shall be deemed to be nil.

(b) the cash reserves of the Group shall be the total sum standing to the credit of the Group in a bank current or deposit account, building society or unit trust, or similarly invested, less £100. If such total sum shall not exceed £100 then the cash reserves of the Group shall be deemed to be nil.

If the option is not exercised at such meeting the withdrawing member shall then use his best endeavours to introduce a new member into the Group and may come to such agreement as he thinks fit with the prospective member as to the terms on which the introduction is to be effected.

If eight weeks shall have passed after written notice of withdrawal has been given to the secretary and the withdrawing member shall not by then have offered his share to the Group then he shall nevertheless be deemed to offer it at the next quorate general meeting thereafter and the Group shall then have the right to exercise its option. If it does not then exercise the option the withdrawing member shall use his best endeavours to introduce a new member as before.

10. *Admission:* A person shall only be admitted as a member of the Group if not less than seven existing members in general meeting so resolve. On the election of a new member the secretary shall notify the same to him in writing, furnish him with a copy of the rules and require him to undertake in writing to abide by them. If the retiring or retired member in whose place the new member is admitted shall not have paid all contributions accrued due from him (as required by Rules 18 and 19) immediately prior to the admission of the new member the latter shall be required to pay such contributions on being given notice to that effect by the secretary. If payment be not made within one month from the date of the secretary's notice the election shall be void unless sufficient cause be shown to the satisfaction of the committee.

11. *Powers:* All matters concerning the conduct and activities of the Group shall be decided by the members in general meeting, save that:
   (a) the committee shall have power without first referring to the members in general meeting to buy any items of equipment not exceeding fifteen pounds (£15.00) in value which they shall consider necessary to further the objects of the Group;
   (b) the decision whether or not to fly or inflate the balloon on any particular occasion shall be in the sole discretion of the pilot in charge for that occasion whose decision shall be final and binding;
   (c) the decision as to which members shall fly on any particular occasion shall be made by the committee as provided for in Rule 14;
   (d) the decisions as to the matters specified in Rules 20 and 21 hereof shall be those of the committee.

12. *Cheques:* The authorised signatories on cheques drawn on the Group's bank account shall be the chairman and the treasurer.

13. *Contracts:* Any two members of the committee together shall be empowered to sign on behalf of the Group any contract or agreement relating in any manner whatsoever to the activities of the Group provided a resolution in favour of such contract or agreement shall first have been passed by the members in general meeting in the manner provided in Rule 7.

14. *Flight Rota:* The secretary shall keep full and proper minutes of all general meetings and shall supply each member with a copy thereof

as soon after each meeting as is reasonably practical. Minutes shall also be kept of all committee meetings.

The secretary shall keep all members as fully informed as possible of proposed flights and together with both other members of the committee shall from time to time prepare a rota with the object so far as possible of giving equal flying opportunities to all members during the course of any twelve-month period.

15. *Treasurer:* The treasurer shall keep the accounts of the Group and shall make up the annual statement of accounts and balance sheet of the Group to 31 October in each year, which shall be printed and circulated amongst the members with the notice of the annual general meeting.

16. *Expulsion:* Any nine members in general meeting shall have power to expel any member who shall offend against the rules of the Group or whose conduct shall in the opinion of the members render him unfit for membership of the Group. For the purposes of this section a quorum shall be nine.

At the quorate general meeting next following the meeting expelling the member he shall offer (or failing which shall be deemed to offer) his share to the Group which shall have the option of purchasing the same on the terms as to payment set out in Rule 9 hereof.

If the option is not then exercised the expelled member shall use his best endeavours to introduce a new member to the Group and may come to such agreement as he thinks fit with the prospective member as to the terms on which the introduction is to be effected.

17. *Dissolution:* If at any general meeting at which not less than nine members are present a resolution for the dissolution of the Group shall be passed by not less than nine members the committee shall thereupon, or at such future date as shall be specified in such resolution, proceed to realise the property of the Group (provided the necessary resolution(s) have first been passed as mentioned in Rule 13) and after the discharge of all liabilities shall divide the same equally among all the members and upon the completion of such division the Group shall be dissolved.

18. *Costs:* All costs of a recurrent nature incurred in carrying out the objects of the Group (such as the premium on a policy of third-party insurance, the cost of repairs and maintenance to the balloon and the cost of storage) but not including those expenses detailed in Rule 19 shall be met by a periodic subscription payable by each member from time to time as the members in general meeting shall decide.

19. *Costs:* The cost of propane used between the start of an inflation and the final landing shall be apportioned equally amongst those flying on any particular occasion at the rate of six pounds (£6.00) per logged flying hour save that the cost apportioned to the pilot in charge shall

be borne out of Group funds provided he is not a member of the Group. Payment shall be made to the treasurer as soon after the flight as possible. The cost of an inflation followed by merely a tethered flight shall be borne out of Group funds.

The cost to each member of travelling to any launching site shall be borne by that member personally, likewise the cost of a retrieval.

20. *Costs:* In the event of there being any dispute as to the division between members of the costs and expenses mentioned above or the amount of any re-imbursement due to any particular member or members the decision of any two members of the committee shall be final and binding upon all members.

21. *Revenue:* Should any payment be made by any person not a member of the Group towards the costs and expenses entailed in any flight or inflation the same shall be dealt with in such manner as the committee shall think fit.

22. *Bank Surplus:* Should there at any time be in the Group's bank account a surplus after taking into account all those items of expenditure accrued due the same shall be dealt with in such manner as the members in general meeting may decide.

23. *Flight by non-Members:* The decision as to whether any non-members shall be allowed to fly on any occasion shall be that of the pilot in charge on that occasion, provided that such non-member must first complete and sign the Group's standard form of release from possible liability. Should a non-member be carried in breach of this proviso the pilot shall indemnify all other members against any costs, claims and expenses that may be incurred by them in connection with an action brought by the non-member.

24. *Damage to Equipment:* Any financial loss suffered by the Group as a result of damage to or loss of the balloon, its road trailer or its associated equipment however caused and whether or not any member shall have been negligent shall be borne equally by all members of the Group.

SUGGESTIONS FOR FURTHER READING

LIGHT AIRCRAFT

The Student Pilot's and Private Pilot's Licence, Civil Air Publication 53: HMSO

Radiotelephony Procedure, Civil Air Publication 46: HMSO

Aviation Law for Applicants for the Private Pilot's Licence, Civil Air Publication 85: HMSO

The BLAC Guide to Navigation for Candidates for the Private Pilot's Licence: British Light Aviation Centre

The BLAC Guide to the IMC Rating: British Light Aviation Centre

The BLAC Manual of Flying and Ground Training: British Light Aviation Centre

Flight Briefing for Pilots: Vols 1–4

Flying Light Aircraft, David Ogilvy

GLIDING

Gliding—A Handbook on Soaring, Derek Piggott

Meteorology for Glider Pilots (also for Balloonists), C. E. Wallington

Meteorology Simplified, J. I. Fell

Go Gliding, Ann Welch and Gabor Denes

BALLOONING

Hot-Air Ballooning, Christine Turnbull

Throw Out Two Hands, Anthony Smith

The Flight of the Small World, Arnold Eiloart and Peter Elstob

PARACHUTING

Sport Parachuting, Charles Shea-Simonds

Alone in the Sky, Mike Reilly

ACKNOWLEDGEMENTS

It would be difficult to name every person who has assisted me in preparing this book, such as those many air-minded enthusiasts who bade me bear in mind that I was getting VIP treatment in my researches at clubs and schools, and so was not called upon to help with less exciting groundwork, or queue impatiently at a glider launch-point, or re-pack a balloon.

Individuals I would, however, like to mention are Radford Barrett and Anthony Marshall, both *Daily Telegraph* colleagues of mine. Ted Barrett applied his editing talent to my text, to considerable advantage, while Tony Marshall, photographer and power pilot, was invaluable with his camera and flying knowledge. Finally, to my sister Margaret who has produced the manuscript from a mass of rewritten notes.

I would also like to thank the following organisations for considerable co-operation and assistance:

The Civil Aviation Authority; National Air Traffic Services; Aeronautical Information Service; Royal Air Force Strike Command; Royal Air Force ATC/CCF; City of London School of Navigation; British Light Aviation Centre; Popular Flying Association; London School of Flying, Elstree; Air Touring Club, Biggin Hill; Rimmer Aviation; Rogers Aviation; Alan Mann Helicopters.

British Gliding Association; London Gliding Club, Dunstable; Lakes Gliding Club, Barrow; The Hang-Gliding Association.

British Balloon & Airship Club; Cameron Balloons; Dante Balloon Group; Europa Balloon School.

British Parachute Association; Peterborough Parachute Centre; The Association of Parascending Clubs.

C. T. Bowring (Insurance) Co Ltd; Slater, Walker Finance Corporation; Lombard North Central; Imperial War Museum; RAF Historical Records.